T0295354

Optimization of Multilayered Radar Absorbing Structures (RAS) using Nature Inspired Algorithm

Optimization of Multilayered Radar Absorbing Structures (RAS) using Nature Inspired Algorithm

Vineetha Joy, Vishal G. Padwal, Hema Singh
and Raveendranath U. Nair

CRC Press
Taylor & Francis Group
Boca Raton London New York

CRC Press is an imprint of the
Taylor & Francis Group, an **informa** business

First edition published 2022
by CRC Press
6000 Broken Sound Parkway NW, Suite 300, Boca Raton, FL 33487-2742

and by CRC Press
2 Park Square, Milton Park, Abingdon, Oxon, OX14 4RN

CRC Press is an imprint of Taylor & Francis Group, LLC

ISBN: 978-0-367-75912-4 (hbk)
ISBN: 978-0-367-75918-6 (pbk)
ISBN: 978-1-003-16456-2 (ebk)

Typeset in Palatino
by Deanta Global Publishing Services, Chennai, India

To

Our Families

Contents

Preface

In recent decades, stealth technology has emerged as a major determinant of defense superiority among global super powers. It plays a crucial role in the combat zone, where swiftness, surprise and instantaneity are the decisive elements for survivability. The supreme goal here is to reduce the radar cross section (RCS) of military vehicles from rival radar systems, thereby allowing the user to conduct surprise military missions. Several techniques like shaping, application of radar absorbing materials (RAM), passive cancellation, active cancellation, etc., can be used to minimize RCS. Among all of them, multilayered radar absorbing structures (RAS) have gained exceptional research interest recently due to their broadband RCS reduction capabilities in combination with reduced weight penalties. The various design parameters in a multilayered RAS model can be tailored to achieve the desired performance over the specified range of frequencies and incident angles. However, when a huge database of potential materials is available, the selection of suitable material for each layer becomes extremely cumbersome. Furthermore, the thickness of each layer has to be appropriately designed so as to get appreciable performance. This is a clear case of an optimization problem where the position of a particular material and its thickness become the parameters to be optimized. Although commercially available software suites provide algorithms (like particle swarm optimization (PSO), genetic algorithm (GA), etc.), for optimization of thickness, they do not have options for optimizing the position of a particular material inside the multilayered RAS configuration. In this regard, this book presents an efficient algorithm, based on particle swarm optimization, for the material selection as well as optimization of thickness of multilayered RAS models considering both normal and oblique incidence cases. It includes a thorough overview of the theoretical background required for the analysis of multilayered RAS as well as the step-by-step procedure used for the implementation of PSO-based optimization algorithm. Further, the accuracy and computational efficiency of the indigenously developed algorithm in comparison with full wave simulation software is also established using suitable validations and case studies. This book will serve as a valuable resource for students, researchers, scientists and engineers involved in the electromagnetic design and development of multilayered radar absorbing structures.

Acknowledgment

First and foremost, we would like to thank God Almighty for being a beacon of hope during the course of this research work.

Further, we would like to thank Shri. Jitendra J. Jadhav, Director, CSIR-National Aerospace Laboratories, Bengaluru, for permission to write this book.

We would also like to acknowledge the valuable suggestions from our colleagues at the Centre for Electromagnetics during the course of writing this book.

But for the concerted support and encouragement of Dr. Gagandeep Singh, Publisher (Engineering), CRC Press, Taylor & Francis Books India Pvt. Ltd., it would not have been possible to bring out this book within such a short span of time.

We are also forever beholden to our family members for their incessant inspiration which supported us to stay on rough tracks.

Vineetha Joy would like to specially thank her husband, Johnu George, for his constant support and encouragement during the course of this work.

Vishal G. Padwal would like to thank his parents for their motivation and support during this tenure.

Hema Singh would like to thank her daughter, Ishita Singh, for her constant cooperation and encouragement during the preparation of the book.

Authors Biography

Vineetha Joy is working as Scientist at Centre for Electromagnetics of CSIR-National Aerospace Laboratories, Bengaluru, India. She obtained B. Tech in Electronics and Communication Engineering with Second Rank from University of Calicut in 2011 and received M. Tech degree with First Rank in RF & Microwave Engineering from Indian Institute of Technology (IIT), Kharagpur, in 2014. She is actively involved in various R&D programs spanning domains such as computational electromagnetics, electromagnetic design and performance analysis of radomes, design and development of broadband radar absorbing structures, hybrid numerical techniques for scattering analysis, and electromagnetic characterization of potential materials for airborne structures. She has authored/co-authored over 100 research publications, including one book, peer-reviewed journal papers, symposium papers, technical documents and test reports.

Vishal G. Padwal obtained his B.E. in Electronics & Tele-communication Engineering from University of Pune, India, and M. E. in Electronics and Tele-communication (Microwave) in 2017 from Department of Electronics & Tele-communication Engineering, Pune Institute of Computer Technology, India. He is currently working as Research Scientist at Society for Applied Microwave Electronics Engineering & Research (SAMEER), Mumbai. In 2018–2019, he was working as Project Assistant at Centre for Electromagnetics of CSIR-National Aerospace Laboratories, Bengaluru, India. His research interest includes EM design and analysis of multilayered radar absorbing structures, optimization techniques, computational electromagnetics, etc. He has co-authored several conference papers, technical documents and test reports.

Dr. Hema Singh is working as Senior Principal Scientist at the Centre for Electromagnetics, National Aerospace Laboratories (CSIR-NAL), Bengaluru, India. She received her PhD degree in Electronics Engineering from IIT-BHU, Varanasi, India, in 2000. For the period 1999–2001, she was Lecturer in Physics at P.G. College, Kashipur, Uttaranchal, India. She was Lecturer in EEE of Birla Institute of Technology & Science (BITS), Pilani, Rajasthan, India, for the period 2001–2004. She joined CSIR-NAL as Scientist in January 2005. Her active areas of research are computational electromagnetics for aerospace applications, EM analysis of propagation in an indoor environment, phased arrays, conformal antennas, radar cross section studies including active RCS reduction. She has contributed to projects sponsored by DRDO on low RCS phased array, active RCS reduction and RAS development, and also in a project sponsored by Boeing USA on EM analysis of RF field build-up within

Boeing 787 Dreamliner. She received Best Woman Scientist Award in CSIR-NAL, Bengaluru for the year 2007–2008 for her contribution in the area of active RCS reduction. Dr. Singh has co-authored 14 books, 2 book chapters, 7 software copyrights, 370 scientific research papers and technical reports.

Dr. Raveendranath U. Nair is working as Senior Principal Scientist & Head at the Centre for Electromagnetics, National Aerospace Laboratories (CSIR-NAL), Bengaluru, India. He holds a PhD degree in Microwave Electronics from MG University, India. He has over 20 years' experience in the field of electromagnetic design and analysis of radomes, design and development of FSS-based structures for airborne platforms, radar cross section studies, design and development of artificially engineered materials, etc. He has contributed significantly to the National Radome Programs, including the Doppler Weather Radar (DWR) radome installed at SHAAR Sreeharikota, Fire Control Radar (FCR) radome for Jaguar aircraft, Astra missile ceramic radome, nosecone radome for Saras aircraft, multiband radomes for TU-142M aircraft, etc. He has authored/co-authored over 200 research publications, including peer-reviewed journal papers, symposium papers, technical reports and two books.

Introduction

Stealth technology is a crucial pre-requisite in the combat zone, where swiftness, surprise and initiative are the decisive elements for survivability. The supreme goal here is to reduce the visibility of military vehicles by shaping, application of radar absorbing materials, passive cancellation, active cancellation, etc. Among all of them, multilayered radar absorbing structures have gained exceptional research interest recently due to their broadband RCS reduction capabilities in combination with reduced weight penalties. The selection of suitable material for each layer and the optimization of the corresponding thickness profile are the two critical factors determining the performance of multilayered RAS. However, when a huge database of potential materials is available, both of these tasks become extremely cumbersome. Although commercially available software suites provide algorithms for optimization of thickness, they do not have options for optimizing the position of a particular material. In this regard, this book presents an efficient algorithm, based on particle swarm optimization, for the material selection as well as optimization of thickness of multilayered RAS models considering both normal as well as oblique incidence cases. It includes a thorough overview of the theoretical background required for the analysis of multilayered RAS as well as the step-by-step procedure for the implementation of PSO-based algorithm. The accuracy and computational efficiency of the indigenously developed code is also clearly established using relevant validations and case studies. This book will serve as a valuable resource for the students, researchers, scientists and engineers involved in the electromagnetic design and development of multilayered radar absorbing structures.

1

Introduction

Low observable technology or stealth technology is a military tactic which covers the whole gamut of methodologies used in electronic warfare to increase the survivability of defense vehicles in battlefields. The predominant aim of stealth technology is to reduce the visibility of an object from rival radar systems. It therefore gives an upper hand to the user by making it more difficult for the enemy to detect the opponent. This in turn allows the user to conduct surprise military missions. The driving goal in the implementation of low observable technology in defense vehicles is the reduction of radar cross section (RCS). Knott *et al.*, in 1985, present a comprehensive overview of the fundamentals of RCS, techniques for prediction/computation of RCS and methods for reduction of RCS. The popular methods used for the reduction of RCS are summarized in the following section.

1.1 Methods for Reduction of Radar Cross Section

There are mainly four approaches that are usually employed to reduce the RCS of military assets. The first technique includes modification of the shape of the military vehicle. Here, the low observable platform is shaped so as to redirect the incident electromagnetic wave in a direction other than directly back to the radar. The second approach employing radar absorbing materials (RAM) uses either novel composites as the base material for the military vehicle or special coatings at identified hotspots. The primary goal of this technique is to absorb, scatter or cancel the incident radar signals so as not to reflect them back to the probing radar. The third approach is called passive cancellation, where a secondary scatterer is introduced to cancel out the reflections from the primary target. The fourth one is the active cancellation scheme, where active transmitters are used to create modified waveforms to cancel the original radar signals reflected from the platform. Another additional approach is the plasma stealth technology, where the incident electromagnetic waves are absorbed using a plasma layer formed with ionized and conductive gas particles.

All the above-mentioned approaches have their own inherent drawbacks. For instance, in the case of a military aircraft, modification of the shape may affect the aerodynamic performance and can cause maneuverability issues. Furthermore, the probability of detection from bistatic radars cannot be overlooked in this case. Passive cancellation schemes are inherently narrowband due to the constraint of quarter-wavelength thickness. On the other hand, active cancellation schemes require complex electronic systems to generate exact waveforms that can cancel out the reflected signals. Considering the case of RAMs, absorbers with a single layer of material have the limitations of narrow frequency band and bulky structure.

1.2 Significance of Multilayered Radar Absorbing Structures (RAS) and Design Challenges

In view of the disadvantages associated with conventional RCS reduction techniques, more attention is being given to the development of multilayered radar absorbing structures, which provide a viable solution for achieving broadband RCS reduction with reduced weight penalty (Singh *et al.*, 2017). The rapid advancements in the field of material science and process engineering have also simplified the fabrication complexities associated with its practical implementation. The various design parameters in a multilayered RAS model (number of material layers, constitutive parameters of each layer, thickness of layers, etc.) can be tailored to achieve the desired performance over the specified range of frequencies and incident angles. The mathematical interpretation/formulation on the reduction of RCS using RAS can be found in Knott *et al.*, 1985; Vinoy and Jha, 1996 and Perini and Cohen, 1993.

However, when a huge database of potential materials is available, the selection of suitable material for each layer becomes extremely cumbersome. Furthermore, the thickness of each layer has to be appropriately designed so as to get appreciable performance. This is a clear case of an optimization problem, where the position of a particular material as well as its thickness becomes the parameters to be optimized. Although commercially available software suites provide algorithms (like particle swarm optimization (PSO), genetic algorithm (GA), etc.) for optimization of thickness, they do not have options for optimizing the position of a particular material inside the multilayered RAS configuration.

The objective of this book is to provide a thorough understanding of the electromagnetic (EM) design, analysis and optimization of multilayered RAS configurations using a nature inspired algorithm. This book consists of five chapters. The full wave formulation used for the computation of overall reflection and transmission coefficients of multilayered dielectric media, for normal as well as oblique incidence, is elaborately described in Chapter 2. The fundamentals of PSO, including a thorough description of its key parameters, are presented in Chapter 3. Further, the step-by-step procedure used for the implementation of PSO-based optimization algorithm in the context of multilayered RAS design is described in detail in Chapter 4, followed by case studies and validations in Chapter 5.

2

EM Wave Propagation in Multilayered Radar Absorbing Structures (RAS)

Multilayered radar absorbing structures (RAS) are generally composed of several material layers stacked on top of one another, resulting in a number of interfaces at which reflection and transmission occur. The theoretical formulation for the computation of overall reflection and transmission coefficients of such configurations is elaborately described in this chapter along with intuitive ray diagrams (Chew, 1995). The formulation clearly takes into account the separate cases of normal as well as oblique incidence.

2.1 Computation of Overall Reflection/Transmission Coefficients in Multilayered RAS for Normal Incidence

The schematic of a multilayered RAS coated over carbon fiber reinforced plastic (CFRP) is shown in Figure 2.1. CFRP is the preferred base material for airborne platforms, and it behaves similarly to metals (constitutive parameters of CFRP are given in the Appendix). In view of this, the bottom layer of all the RAS models considered in this book is fixed as CFRP with a thickness of 3.0 mm. Multilayered RAS consist of N layers, with the mth layer having specific thickness (t_m), complex permeability $(\mu_m = \mu'_m - j\mu''_m)$ and complex permittivity $(\varepsilon_m = \varepsilon'_m - j\varepsilon''_m)$. The material suitable for each layer has to be chosen appropriately from a pre-defined database so as to have reduced reflection as well as transmission. The main objective is to obtain maximum power absorption from the designed RAS. Furthermore, the thickness profile has to be optimized for achieving broadband characteristics.

A plane wave incident normally along the Z-axis, as shown in Figure 2.2, is considered for the electromagnetic (EM) analysis of the multilayered RAS configuration. The material layers are assumed to be extending infinitely in the XY plane. Here, multiple reflections occur at each interface of the RAS for both down-going and up-coming waves. The Fresnel reflection coefficients (R) need to be computed for each layer and these coefficients will be used for the computation of recursive reflection coefficients at every interface.

The intrinsic reflection and transmission coefficients for a particular interface have been formulated by Fresnel in 1823 (Chew, 1995) and these

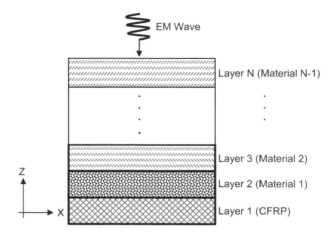

FIGURE 2.1
Schematic of multilayered radar absorbing structure.

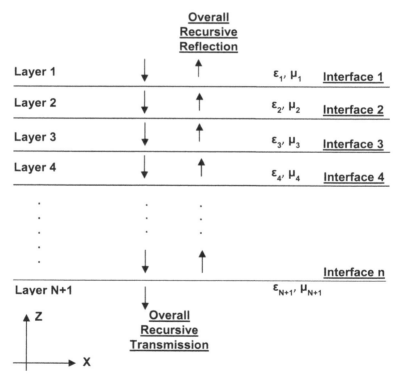

FIGURE 2.2
Reflection and transmission through multilayered RAS for normal incidence.

coefficients have been named after him as Fresnel coefficients. For the case of normal incidence, the wave remains normal to every interface, as shown in Figure 2.2, as it propagates through the constituent layers. The Fresnel reflection and transmission coefficients at the interface between i^{th} and $(i + 1)^{th}$ layer, in terms of permittivity (ε), permeability (μ) and phase constant (k), can be calculated using Equations (2.1) to (2.4). In the case of normal incidence ($\theta_i = \theta_t = 0$), the coefficients obtained for transverse electric (TE) and transverse magnetic (TM) polarizations will be the same.

For TE polarization (Balanis, 2012),

$$R^{TE}_{i,i+1} = \frac{\mu_{i+1} k_i \cos\theta_i - \mu_i k_{i+1} \cos\theta_t}{\mu_{i+1} k_i \cos\theta_i + \mu_i k_{i+1} \cos\theta_t} \tag{2.1}$$

$$T^{TE}_{i,i+1} = \frac{2\mu_{i+1} k_i \cos\theta_i}{\mu_{i+1} k_i \cos\theta_i + \mu_i k_{i+1} \cos\theta_t} \tag{2.2}$$

For TM polarization (Balanis, 2012),

$$R^{TM}_{i,i+1} = \frac{\varepsilon_{i+1} k_i \cos\theta_t - \varepsilon_i k_{i+1} \cos\theta_i}{\varepsilon_{i+1} k_i \cos\theta_t + \varepsilon_i k_{i+1} \cos\theta_i} \tag{2.3}$$

$$T^{TM}_{i,i+1} = \frac{2\varepsilon_{i+1} k_i \cos\theta_i}{\varepsilon_{i+1} k_i \cos\theta_t + \varepsilon_i k_{i+1} \cos\theta_i} \tag{2.4}$$

where $R_{i, i+1}$ and $T_{i, i+1}$ are the Fresnel reflection and transmission coefficients, respectively. The subscripts i and $i+1$ denote the particular material layer under consideration.

For normal incidence,

$$R^{TE}_{i,i+1} = R^{TM}_{i,i+1} \text{ and } T^{TE}_{i,i+1} = T^{TM}_{i,i+1}$$

Once the Fresnel coefficients are computed, the generalized formula for the recursive reflection coefficient ($\tilde{R}$) at the interface between layer i and layer $i+1$ can be expressed as (Chew, 1995),

$$\tilde{R}_{i,i+1} = \frac{R_{i,i+1} + \tilde{R}_{i+1,i+2} e^{2jk_{i+1,z}(t_{i+1}-t_i)}}{1 + R_{i,i+1}\tilde{R}_{i+1,i+2} e^{2jk_{i+1,z}(t_{i+1}-t_i)}} \tag{2.5}$$

where t denotes the thickness of the material layer under consideration. If the EM wave is incident on N^{th} layer, the overall recursive reflection coefficient corresponds to the total reflection coming from N^{th} layer, which in turn includes the effect of reflections from all the layers beneath it. This recursive effect is illustrated in Figure 2.3. Here, $\tilde{R}_{12}$ will denote the overall recursive reflection coefficient corresponding to the multilayered RAS.

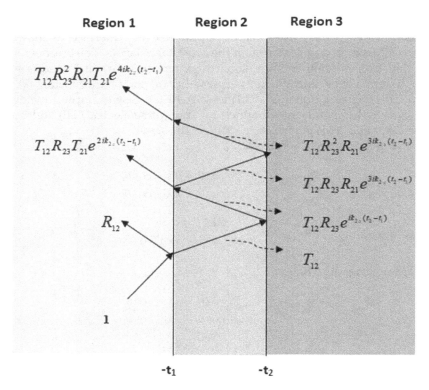

FIGURE 2.3
Illustration of multiple reflections in three-layered media.

Similarly, the overall recursive transmission coefficient of the multilayered RAS can be defined as,

$$\tilde{T}_{1N} = \prod_{p=1}^{N-1} e^{jk_{pz}\left(t_p - t_{p-1}\right)} S_{p,p+1} \tag{2.6}$$

where,

$$S_{p,p+1} = \frac{T_{p,p+1}}{1 - R_{p+1,p}\tilde{R}_{p+1,p+2}e^{2jk_{p+1z}(t_{p+1}-t_p)}}$$

2.2 Computation of Overall Reflection/Transmission Coefficients in Multilayered RAS for Oblique Incidence

The reflection and transmission of EM waves through multilayered RAS for oblique incident angles is shown in Figure 2.4. The formulation presented in

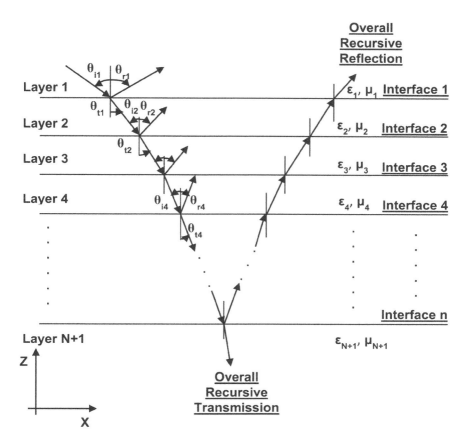

FIGURE 2.4
Reflection and transmission through multilayered RAS for oblique incidence.

the previous section has to be revised to include the effect of non-zero angle of incidence and the angle of incidence at each interface needs to be evaluated using Snell's law of refraction.

Snell's law of refraction can be stated as,

$$k_i \sin\theta_{iN} = k_{i+1}\sin\theta_{tN} \tag{2.7}$$

where θ_i and θ_t are the angles made by the incident ray and the refracted ray, respectively, with the normal to the interface N.

For interface 1,

$$k_1 \sin\theta_{i1} = k_2 \sin\theta_{t1}$$

$$\sin\theta_{t1} = \frac{k_1}{k_2}\sin\theta_{i1} \tag{2.8}$$

For interface 2,

$$k_2 \sin \theta_{i2} = k_3 \sin \theta_{t2}$$

$$\sin \theta_{t2} = \frac{k_2}{k_3} \sin \theta_{i2} \tag{2.9}$$

It is clear from Figure 2.4 that $\theta_{t1} = \theta_{i2} = \theta_{r2}$.
Replacing θ_{i2} in (2.9) by θ_{t1} and substituting for $\sin \theta_{t1}$ gives,

$$\sin \theta_{t2} = \frac{k_2}{k_3} \sin \theta_{t1}$$

$$\sin \theta_{t2} = \frac{k_2}{k_3} \times \frac{k_1}{k_2} \sin \theta_{i1} = \frac{k_1}{k_3} \sin \theta_{i1}$$

Therefore, a general expression can be written as,

$$\sin \theta_{tN} = \frac{k_1}{k_{N+1}} \sin \theta_{i1} \tag{2.10}$$

Once the incident angles at every interface are found, the Fresnel reflection and transmission coefficients at each interface have to be evaluated for both TE as well TM polarizations. The theoretical derivation for these coefficients is discussed in the following sections.

2.2.1 Transverse Electric (TE) Polarization

In transverse electric or perpendicular polarization, the electric field vector is perpendicular to the plane of incidence. The directions of electric field vector, magnetic field vector and propagation vector for incident wave, reflected wave and transmitted wave in accordance with Poynting's theorem are illustrated in Figure 2.5.

The field equations for incident wave in N^{th} layer can be written as (Pozar, 1998),

$$H_{ix} = \frac{E_0}{\eta_N} \cos \theta_i e^{-jk_N(x \sin \theta_i)} \tag{2.11}$$

$$H_{iz} = \frac{E_0}{\eta_N} \sin \theta_i e^{-jk_N(-z \cos \theta_i)} \tag{2.12}$$

$$E_{iy} = E_0 e^{-jk_N(x \sin \theta_i - z \cos \theta_i)} \tag{2.13}$$

where E_0 and η_N denote the amplitude of incident electric field and wave impedance of Nth layer, respectively.

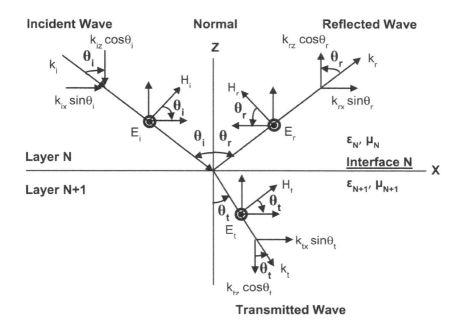

FIGURE 2.5
Reflection and transmission at N^{th} interface of a multilayered RAS for obliquely incident TE wave.

Similarly, the equations for reflected and transmitted fields can be written as (Pozar, 1998),

$$H_{rx} = \frac{-R_{TE}E_0}{\eta_N} \cos\theta_r e^{-jk_N(x\sin\theta_r)} \tag{2.14}$$

$$H_{rz} = \frac{R_{TE}E_0}{\eta_N} \sin\theta_r e^{-jk_N(z\cos\theta_r)} \tag{2.15}$$

$$E_{ry} = R_{TE}\,E_0 e^{-jk_N(x\sin\theta_r + z\cos\theta_r)} \tag{2.16}$$

$$H_{tx} = \frac{T_{TE}E_0}{\eta_{N+1}} \cos\theta_t e^{-jk_{N+1}(x\sin\theta_t)} \tag{2.17}$$

$$H_{tz} = \frac{T_{TE}E_0}{\eta_{N+1}} \sin\theta_t e^{-jk_{N+1}(-z\cos\theta_t)} \tag{2.18}$$

$$E_{ty} = T_{TE}E_0 e^{-jk_{N+1}(x\sin\theta_t - z\cos\theta_t)} \tag{2.19}$$

where R_{TE} and T_{TE} denote the Fresnel reflection and transmission coefficients of Nth interface for TE polarization, respectively.

Applying tangential boundary conditions at the interface for E_y and H_x gives,

$$E_0 e^{-jk_N (x \sin \theta_i)} \hat{y} + E_0 e^{-jk_N (x \sin \theta_r)} R_{TE} \hat{y} = E_0 e^{-jk_{N+1}(x \sin \theta_t)} T_{TE} \hat{y} \tag{2.20}$$

$$\frac{E_0}{\eta_N} \cos \theta_i e^{-jk_N (x \sin \theta_i)} \hat{x} + \frac{(-E_0)}{\eta_N} \cos \theta_r e^{-jk_N (x \sin \theta_r)} R_{TE} \hat{x}$$

$$= \frac{E_0}{\eta_{N+1}} \cos \theta_t e^{-jk_{N+1}(x \sin \theta_t)} T_{TE} \hat{x} \tag{2.21}$$

Equations (2.20) and (2.21) simplify to Equations (2.22) and (2.23), respectively, on applying Snell's law ($\theta_i = \theta_r$).

$$1 + R_{TE} = T_{TE} \tag{2.22}$$

$$\frac{\cos \theta_i}{\eta_N} (1 - R_{TE}) = \frac{\cos \theta_t}{\eta_{N+1}} T_{TE} \tag{2.23}$$

The Fresnel reflection and transmission coefficients (Pozar, 1998; Balanis, 2012) of Nth interface for TE polarization can be found by solving (2.22) and (2.23) and they can be written as,

$$R_{TE} = \frac{\eta_{N+1} \cos \theta_{iN} - \eta_N \cos \theta_{tN}}{\eta_{N+1} \cos \theta_{iN} + \eta_N \cos \theta_{tN}} \tag{2.24}$$

$$T_{TE} = \frac{2 \eta_{N+1} \cos \theta_{iN}}{\eta_{N+1} \cos \theta_{iN} + \eta_N \cos \theta_{tN}} \tag{2.25}$$

2.2.2 Transverse Magnetic (TM) Polarization

In transverse magnetic or parallel polarization, the electric field vector is parallel to the plane of incidence, whereas the magnetic field vector is perpendicular to it. The directions of electric field vector, magnetic field vector and propagation constant for incident wave, reflected wave and transmitted wave in accordance with Poynting's theorem are illustrated in Figure 2.6.

The field equations for incident wave in N^{th} layer can be written as,

$$E_{ix} = -E_0 \cos \theta_i e^{-jk_N (x \sin \theta_i)} \tag{2.26}$$

$$E_{iz} = -E_0 \sin \theta_i e^{-jk_N (-z \cos \theta_i)} \tag{2.27}$$

$$H_{iy} = \frac{E_0}{\eta_N} e^{-jk_N (x \sin \theta_i - z \cos \theta_i)} \tag{2.28}$$

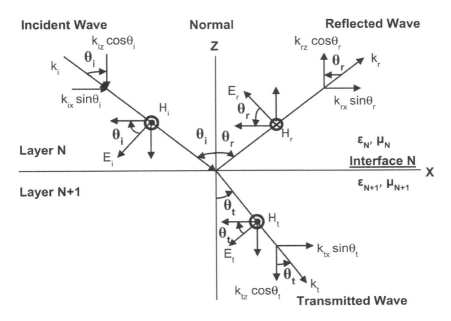

FIGURE 2.6
Reflection and transmission at N^{th} interface of a multilayered RAS for obliquely incident TM wave.

Similarly, the equations for reflected and transmitted fields can be written as (Pozar, 1998),

$$E_{rx} = -R_{TM}E_0 \cos\theta_r e^{-jk_N(x\sin\theta_r)} \tag{2.29}$$

$$E_{rz} = R_{TM}E_0 \sin\theta_r e^{-jk_N(z\cos\theta_r)} \tag{2.30}$$

$$H_{ry} = \frac{-R_{TM}E_0}{\eta_N} e^{-jk_N(x\sin\theta_r + z\cos\theta_r)} \tag{2.31}$$

$$E_{tx} = -T_{TM}E_0 \cos\theta_t e^{-jk_{N+1}(x\sin\theta_t)} \tag{2.32}$$

$$E_{tz} = -T_{TM}E_0 \sin\theta_t e^{-jk_{N+1}(-z\cos\theta_t)} \tag{2.33}$$

$$H_{ty} = \frac{T_{TM}E_0}{\eta_{N+1}} e^{-jk_{N+1}(x\sin\theta_t - z\cos\theta_t)} \tag{2.34}$$

where R_{TM} and T_{TM} denote the Fresnel reflection and transmission coefficients of Nth interface for TM polarization, respectively.

Applying tangential boundary conditions at the interface for E_x and H_y gives,

$$-E_0 \cos\theta_i e^{-jK_N(x\sin\theta_i)}\hat{x} - E_0 \cos\theta_r e^{-jK_N(x\sin\theta_r)}\Gamma\hat{x}$$
$$= -E_0 \cos\theta_t e^{-jK_{N+1}(x\sin\theta_t)}T\hat{x} \tag{2.35}$$

$$\frac{E_0}{\eta_N} e^{-jk_N(x\sin\theta_i - z\cos\theta_i)}\hat{y} - \frac{E_0}{\eta_N} e^{-jk_N(x\sin\theta_r + z\cos\theta_r)}R_{TM}\hat{y}$$
$$= \frac{E_0}{\eta_{N+1}} e^{-jk_{N+1}(x\sin\theta_t - z\cos\theta_t)}T_{TM}\hat{y} \tag{2.36}$$

Equations (2.35) and (2.36) simplify to Equations (2.37) and (2.38), respectively, on applying Snell's law ($\theta_i = \theta_r$).

$$\cos\theta_i(1 + R_{TM}) = \cos\theta_t T_{TM} \tag{2.37}$$

$$\frac{1}{\eta_N}(1 - R_{TM}) = \frac{T_{TM}}{\eta_{N+1}} \tag{2.38}$$

The Fresnel reflection and transmission coefficients (Pozar, 1998; Balanis, 2012) of Nth interface for TM polarization can be found by solving (2.37) and (2.38) and they can be written as,

$$R_{TM} = \frac{\eta_{N+1}\cos\theta_{tN} - \eta_N\cos\theta_{iN}}{\eta_{N+1}\cos\theta_{tN} + \eta_N\cos\theta_{iN}} \tag{2.39}$$

$$T_{TM} = \frac{2\eta_{N+1}\cos\theta_{iN}}{\eta_{N+1}\cos\theta_{tN} + \eta_N\cos\theta_{iN}} \tag{2.40}$$

For parallel polarization, there exists a special angle of incidence called Brewster's angle (θ_b) at which the reflection coefficient becomes zero, i.e., at $\theta_i = \theta_b$,

$$\eta_{N+1}\cos\theta_{tN} = \eta_N\cos\theta_{bN} \tag{2.41}$$

This in turn reduces to,

$$\frac{\eta_{N+1}}{\eta_N} = \frac{\sqrt{1 - \sin^2\theta_{bN}}}{\sqrt{1 - \sin^2\theta_{tN}}}$$

Substituting for $\sin\theta_{tN} = \dfrac{k_N}{k_{N+1}}\sin\theta_b$ and $\eta_N = k_N / (w\sqrt{\varepsilon_N})$ gives,

$$\frac{\eta_{N+1}^2}{\eta_N^2} = \frac{1-\sin^2\theta_b}{1-\dfrac{k_N^2}{k_{N+1}^2}\sin^2\theta_b} \rightarrow \sin\theta_b = \frac{1}{\sqrt{1+\dfrac{\varepsilon_N}{\varepsilon_{N+1}}}} \tag{2.42}$$

Once the Fresnel reflection and transmission coefficients for both polarizations are calculated for each and every interface, the recursive reflection coefficient at the interface between layer i and layer $i+1$ and the overall recursive transmission coefficient can be expressed as,

$$\tilde{R}_{i,i+1} = \frac{R_{i,i+1} + \tilde{R}_{i+1,i+2}e^{2jk_{i+1,z}(t_{i+1}-t_i)\cos\theta_{ti}}}{1+R_{i,i+1}\tilde{R}_{i+1,i+2}e^{2jk_{i+1,z}(t_{i+1}-t_i)\cos\theta_{ti}}} \tag{2.43}$$

$$\tilde{T}_{1N} = \prod_{p=1}^{N-1} e^{jk_{pz}(t_p-t_{p-1})\cos\theta_{ti}}S_{p,p+1} \tag{2.44}$$

where,

$$S_{p,p+1} = \frac{T_{p,p+1}}{1-R_{p+1,p}\tilde{R}_{p+1,p+2}e^{2jk_{p+1z}(t_{p+1}-t_p)\cos\theta_{ti}}}$$

Here, R and T denote the Fresnel reflection and transmission coefficients for oblique incidence at a particular interface. Considering the polarization of the incident wave, the corresponding Fresnel coefficients, given in Section 2.2.1 and Section 2.2.2, have to be appropriately substituted in Equations (2.43) and (2.44).

Once the overall recursive reflection ($\tilde{R}_{12}$) and transmission ($\tilde{T}_{1N}$) coefficients are computed, the percentage of power absorbed by the multilayered RAS can be evaluated as,

$$Absorbed\ power\ (in\ \%) = (1-\left|\tilde{T}_{1N}\right|^2 - \left|\tilde{R}_{12}\right|^2)\times 100$$

The goal of optimization has to be set so as to achieve a percentage of power absorbed greater than 90% throughout the frequency range of operation with the total thickness of RAS within user-defined limits.

3

Application of Nature Inspired Algorithms for Optimization Problems

Nature inspired algorithms like genetic algorithm (GA), swarm intelligence-based algorithms, differential evolution, central force algorithm, wind driven optimization technique, etc., present efficient approaches for the optimization of practical engineering problems using the wisdom available in nature. A brief review of these algorithms, in the context of radar absorbing structures, is presented in this chapter. Amongst all of them, particle swarm optimization, a computationally brilliant swarm intelligence-based algorithm, has been used in this book for the optimization of multilayered radar absorbing structures. The underlying physical interpretations, key parameters involved and the methodology for the application of particle swarm optimization in engineering problems are also elaborately described in this chapter.

3.1 Nature Inspired Algorithms for Optimization of RAS

The electromagnetic design of radar absorbing structures is a multi-objective optimization problem where the scattered energy is to be minimized over a broad range of frequencies and incident angles, at the same time ensuring reduced thickness and weight penalty. It involves several crucial factors like choice of constitutive parameters of materials, number of layers, thickness of individual layers, range of frequency, range of incident angles, type of wave polarization, etc. An electromagnetic engineer needs to choose various design parameters by making intelligent trade-offs between conflicting objectives while keeping the absorber physically realizable and structurally stable. With the availability of a myriad of potential absorbing materials and the rapid advancements in the field of fabrication technologies, this is a herculean task to accomplish with traditional optimization algorithms and brute force approaches.

In this direction, recent decades have been witnessing a surge in the application of nature inspired algorithms for the optimization of RAS. GA has been used for the optimization of planar textured absorbers (Cui *et al.*, 2006), artificial magnetic conductor (AMC) based ultra-thin absorbers (Kern *et al.*, 2003), FSS-based absorbing dielectric composites (Chakravarty *et al.*, 2002), multilayered microwave absorbers (Jiang *et al.*, 2009; Michielssen *et al.*, 1993), etc. However, GA is comparatively difficult to implement due to its innate complexity and prolonged computation time (Goudos *et al.*, 2008; Hassan *et al.*, 2005). Central force optimization (CFO), a novel algorithm based on gravitational kinematics, has been used for the design of multilayered RAS for normally incident plane wave in a pre-defined frequency range (Asi *et al.*, 2010). The same problem has also been attempted using artificial bee colony algorithm (Yigit *et al.*, 2019; Toktas *et al.*, 2018), self-adaptive differential evolution algorithm (DE) (Goudos *et al.*, 2008) and wind driven optimization approach (Ranjan *et al.*, 2017).

Furthermore, particle swarm optimization, an efficient algorithm based on the social behavior of swarms, has been widely used for the optimization of multilayered RAS (Roy *et al.*, 2015; Goudos *et al.*, 2006; Chamaani *et al.*, 2008). Moreover, PSO has been found to have a similar efficiency of GA in the case of multi-dimensional multi-objective problems, but with significantly reduced computational complexity and high ease of implementation (Rania Hassan *et al.*, 2005). Therefore, PSO has been used in this book for the optimization of multilayered RAS.

3.2 Overview of Particle Swarm Optimization

Particle swarm optimization is a robust stochastic evolutionary optimization technique developed by J. Kennedy and R. C. Eberhart (Roy *et al.*, 2015; Robinson *et al.*, 2004). This optimization concept is based on the behavior of bees as they are searching for food. The intelligence used by the swarm of bees can be correlated to real-time problems. Let us consider a swarm of bees in a garden. The goal of all bees is to find the location with maximum density of flowers so that they can collect maximum honey. Initially, all bees start flying from random locations in random directions. Each bee can remember not only the location where it came across the highest density of flowers but also the location of maximum flower density encountered by other bees in the swarm. The bee accelerates in a direction which takes into account the location where it personally came across the maximum number of flowers as well as the best location communicated by other bees.

The trajectory of the bee will depend on whether nostalgia or social influence predominates the decision-making process. As the bee moves along, it might find a new location with higher density of flowers than it had found previously. It would then alter its path according to this newly encountered best location as well as *w.r.t.* the best location encountered so far by the swarm. Further, it may occur that one bee finds a new place with more flowers than had been found by any other bee in the swarm. In such a situation, the entire swarm would be attracted toward this newly found best location in addition to their own personal best location. The swarm explores the entire field in this manner, where they might overfly locations of highest density and get pulled back toward them later. They continuously check the places they fly over against previously found locations of maximum concentration, hoping to reach the place with the absolute highest density of flowers. Finally, the bee reaches the place in the field with the maximum number of flowers. Also, the entire swarm reaches this location soon (Robinson *et al.*, 2004).

3.3 Key Terminology in PSO Algorithm

The key terms used while implementing the PSO algorithm are explained below (Robinson *et al.*, 2004):

a) *Particle*: Each individual bee from the swarm is called a particle. All the particles work on the same principle as mentioned in the previous section. They consistently check their current location with the personal as well as the overall best locations.

b) *Position*: The location of a bee in the garden is referred to as its position, which in turn would be a set of co-ordinates. This idea can be extended to the real-time problems, where each position corresponds to a potential solution to the problem at hand, i.e., the position refers to the set of parameters to be optimized. Therefore, the dimension of position depends on the number of parameters to be optimized.

c) *Fitness function*: It evaluates the quality of a particular position. The fitness function takes in the position of a particular particle and returns back a value which represents the degree of goodness of that position. In the case of a swarm of bees, the concentration of flowers at a particular position is the value of the fitness function at that particular position. Higher the concentration, better would be the position.

d) *Personal best* (*Pbest*): The location with the best value for the fitness function personally encountered by the bee is referred to as its personal best (*Pbest*). *Pbest* depends on the trajectory followed by the bee in its pursuit of the best location. The bee compares the fitness function evaluated at its current location to that given by the *Pbest* location encountered so far. On comparison, if the current location has a better value for the fitness function, *Pbest* would be replaced with its current location.

e) *Overall best/global best (Gbest)*: The location with the best value of the fitness function encountered so far by the swarm of bees is called *Gbest*. At each position along the trajectory, the bee compares the fitness function evaluated at its current location to that given by the *Gbest* location. If found to be better, then *Gbest* would be replaced with the bee's current location.

3.4 Description of Generalized PSO Algorithm

Once the user is familiar with the key terms as explained in the previous section, PSO algorithm can be described (Robinson *et al.*, 2004) using a step-by-step procedure as given below:

Step#1: *Definition of solution space*

This step includes the selection of suitable parameters to be optimized and the respective ranges in which they are allowed to vary. The maximum and minimum values of each dimension (as parameters) have to be clearly identified.

Step#2: *Definition of fitness function*

The fitness function has to be intelligently formulated so as to accurately represent the quality of a position. This function should clearly represent the goodness of a particular position in a single value. The fitness function and solution space will vary with the problem under consideration.

Step#3: *Initialization of random positions and velocities*

In the pursuit of an optimal solution, every particle starts from a random location and moves initially with a velocity random in both magnitude and direction. These random parameters have to be initialized. At this step, the value of fitness function corresponding to the *Pbest* of each particle would be calculated using its respective initial position and the value of fitness function corresponding to *Gbest* would then be chosen from amongst all the available *Pbest* values.

Step#4: *Movement of particles in the solution space*

The particles must now be moved through the solution space for which the algorithm considers each particle, moves it by a small amount and cycles it through the entire swarm. The steps that will be applied on each particle individually are given below:

i. Evaluation of fitness function for the current position of a particle and comparison with the values at its *Pbest* and *Gbest*:

If the value of fitness function is found to be better than that given by the previous *Pbest* and current *Gbest*, the appropriate positions are replaced with the current position.

ii. Updating the velocity of the particle *w.r.t.* the relative positions of *Pbest* and *Gbest*:

The general equation can be expressed as (Roy *et al.*, 2015),

$$V_{iD}^{j} = q \times V_{iD}^{j-1} + k_1 \times rand1_{iD}^{j}() \times \left(pbest_{iD}^{j-1} - X_{iD}^{j-1}\right)$$
$$+ k_2 \times rand2_{iD}^{j-1}() \times \left(gbest_{D}^{j-1} - X_{iD}^{j-1}\right) \tag{3.1}$$

where, *i*: particle number; *j*: iteration number; V_{iD}: velocity of the particle in the Dth dimension; X_{iD}: particle's co-ordinate in the Dth dimension; *q*: inertial weight; k_1, k_2: acceleration constants which determine the relative pull of *Gbest* and *Pbest*; *rand1, rand2*: random number functions which generate uniform random numbers between 0 and 1

iii. Movement of particle using the updated velocity:

The position can be updated as,

$$X_{iD}^{j} = X_{iD}^{j-1} + V_{iD}^{j} \tag{3.2}$$

Step#5: Once the above-mentioned steps are completed for all the particles, the process has to be repeated from Step 4. This continues until a particular condition is satisfied or when it completes a predefined number of iterations. The movement of particles within the solution space for a generalized RAS problem is shown in Figure 3.1.

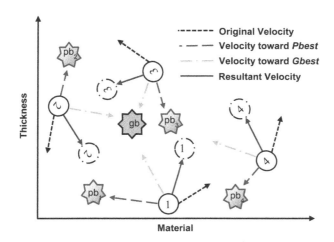

FIGURE 3.1
Movement of four particles (1, 2, 3 and 4) within a 2D solution space (pb: *Pbest*; gb: *Gbest*).

The selection of various constants used in equation (3.1) is also critical for the performance of the optimization algorithm. The particle should explore the complete solution space and at the same time should not go beyond the solution space. The criteria used for the selection of values for various constants are elaborately described in Robinson *et al.* (2004).

As mentioned above, the particles should be confined within the solution space at any point of the algorithm. For achieving this, three boundary conditions are usually suggested and they are given below:

(i) *Absorbing walls*

Under this boundary condition, the particle going out of bounds is pulled back toward the available solution space by making its velocity zero, as shown in Figure 3.2a. When a particle reaches the boundary of a particular dimension, its velocity in that dimension will be made equal to zero.

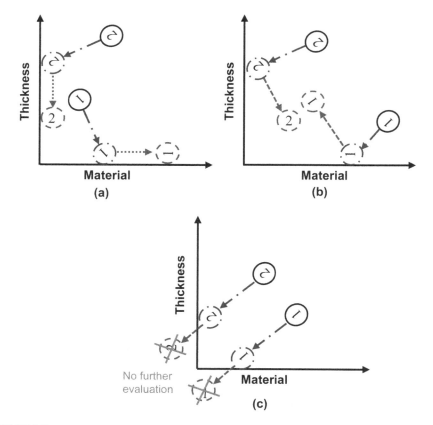

FIGURE 3.2
Boundary conditions used in PSO algorithm. (a) Absorbing walls, (b) Reflecting walls and (c) Invisible walls.

(ii) *Reflecting walls*

Here, the particle going out of bounds is reflected back toward the solution space by changing the sign of the velocity in that dimension. This is illustrated in Figure 3.2b.

(iii) *Invisible walls*

In this case, the particles reaching the boundaries are allowed to go beyond the solution space, as shown in Figure 3.2c, and these particles will no longer be considered in the algorithm.

4

PSO-based Algorithm for the EM Design
and Optimization of Multilayered RAS

The algorithm for PSO in the context of design and optimization of multi-layered RAS is presented in this chapter. The key terminology as well as the implementation methodology is elaborately described against this backdrop. A detailed step-by-step procedure, along with a comprehensive flowchart, is also included to help the reader implement the algorithm with ease.

4.1 Methodology of PSO for the Optimal Design of Multilayered RAS

The general algorithm for PSO, as explained in Chapter 3, can be adapted appropriately to achieve an optimal design for multilayered RAS configurations. In the design of multilayered RAS configurations, the materials to be assigned for each layer have to be chosen judiciously from a pre-defined database. Furthermore, the thickness profile for a particular combination of materials has to be optimized for achieving broadband absorption. This scenario presents an optimization problem where the position of a particular material in the multilayered model and its thickness become the parameters to be optimized. The general PSO algorithm can now be correlated to this design problem. Some of the important terms used in PSO terminology *w.r.t.* the optimization of multilayered RAS are described below:

(a) *Particle*: Here, a particle refers to a potential solution to the multilayered RAS design problem.

(b) *Position*: Each particle can be identified by a position, which in this case would be a set of materials and corresponding thicknesses which together make up the multilayered RAS configuration.

(c) *Fitness*: The value of the fitness function quantifies the quality of a particular position. In this particular design problem, a position can be considered good if the RAS model made up of the set of materials and corresponding thicknesses representing that particular position gives minimum reflection as well as transmission over a broad range of frequencies and incident angles. Therefore, the summation of the overall recursive reflection and transmission coefficients for the particular multilayered RAS model, over a range of frequencies and incident angles, can give a single value which represents the fitness of that position. The lower the value of the fitness function, the better would be the position.

The meaning of *Pbest* and *Gbest* remains the same, irrespective of the problem.

The various steps in the general algorithm for PSO, adapted for the multilayered RAS design problem with N discrete material layers, are described below:

Step#1: *Definition of solution space*

In this case, the solution space is defined by the maximum and minimum values of the thickness corresponding to each layer as well as the number of materials in the pre-defined database (Roy *et al.*, 2015). Therefore, optimization will be performed over this prescribed range of values. Here, the minimum and maximum value for a particular dimension is denoted by X_{min}^D and X_{max}^D, respectively. Further, the minimum and maximum value of the velocity along the particular dimension is denoted by V_{min}^D and V_{max}^D, respectively. For the multilayered RAS design problem, the maximum velocity along a particular dimension is set to the maximum value of that particular dimension, i.e., $V_{max}^D = X_{max}^D$ and the minimum velocity V_{min}^D is set to $-V_{max}^D$.

Step#2: *Definition of fitness function*

As explained at the beginning of this section, the fitness function (F) can be written as,

$$F = \sum_{i}^{n\_f} \sum_{j}^{n\_a} \left(\left| \tilde{R} \right|^n + \left| \tilde{T} \right|^n \right) \tag{4.1}$$

where n_f is the number of frequency points and n_a is the number of incident angles over which the RAS is to have optimum performance. $\left| \tilde{R} \right|$ and $\left| \tilde{T} \right|$ correspond to the magnitude of overall recursive reflection and transmission coefficients, respectively, at i^{th} frequency and j^{th} incident angle. The coefficients corresponding to TE and TM polarization have to be considered depending on the polarization of incident plane wave.

For higher values of n, the frequencies with lower reflection and transmission coefficient will have less importance in the fitness function since both the coefficients are always lesser than one. This aids in moving the focus of optimization to those frequencies with higher reflection/transmission coefficient. In the studies included in this book, n is assigned a value of 2.

Step#3: *Initialization of random particle location and velocities*

Here, the initial position as well as the initial velocity for each particle has to be generated randomly. For an N layered problem, the position of a particle will have a dimension of $2N$, which includes N materials as well as N thicknesses.

Step#4 and Step#5 remains the same as described in Chapter 3.

4.2 Algorithm for the Optimization of Multilayered RAS Using PSO

The step-by-step procedure for the implementation of PSO-based algorithm for the optimal design of an N layered RAS over CFRP is presented in this section. A comprehensive flowchart and elaborate description of all variables and equations are also included here to aid the reader in realizing the algorithm using any programming language. FORTRAN has been used in this book for the same. The detailed procedure is given below:

Step#1: *Specification of user-defined inputs*

The values for number of layers (N), number of particles (P), number of materials (M), number of iterations (*Iteration*), number of frequency points (*FR_P*), maximum and minimum values for the thickness of each layer, etc., have to be taken as user-defined inputs. The maximum and minimum values for the thickness of each layer as well the maximum and minimum values for the material number corresponding to each layer have to be then stored in matrices *DIM_max* and *DIM_min*, respectively. Here, the first element of each row corresponds to the dimension number; t and m correspond to thickness and material number, respectively, for a particular layer; and the subscript i denotes the dimension number. In this case, the maximum value for material number corresponds to the number of materials in the predefined database, i.e., $m_{N+1\max} = m_{N+2\max} = \cdots\cdots, = m_{2N\max} = M$ and the minimum value for material number corresponds to 1, i.e., $m_{N+1\min} = m_{N+2\min} = \cdots\cdots, = m_{2N\min} = 1$. The formats for both the matrices are given below:

$$DIM\_\max = \begin{bmatrix} 1 & t_{1\max} \\ 2 & t_{2\max} \\ \vdots & \vdots \\ N & t_{N\max} \\ N+1 & m_{N+1\max} \\ N+2 & m_{N+2\max} \\ \vdots & \vdots \\ 2N & m_{2N\max} \end{bmatrix} \quad DIM\_\min = \begin{bmatrix} 1 & t_{1\min} \\ 2 & t_{2\min} \\ \vdots & \vdots \\ N & t_{N\min} \\ N+1 & m_{N+1\min} \\ N+2 & m_{N+2\min} \\ \vdots & \vdots \\ 2N & m_{2N\min} \end{bmatrix}$$

Step#2: *Definition of boundaries for velocities*

The maximum and minimum values of velocity for all dimensions (in the form of matrix *VEL_max* and *VEL_min*) are then defined as,

$$VEL\_\max = DIM\_\max$$

$$VEL\_\min = -VEL\_\max$$

Step#3: *Generation of random positions and velocities*

The initial positions of all the particles as well their initial velocities have to be randomly initialized and stored in suitable arrays. The matrix (*POP*) containing the initial positions of all the particles can be written as,

$$POP = \begin{bmatrix} 1 & t_{11} & t_{12} & t_{13} & \cdots & t_{1N} & m_{11} & m_{12} & m_{13} & \cdots & m_{1N} \\ 2 & t_{21} & t_{22} & t_{23} & \cdots & t_{2N} & m_{21} & m_{22} & m_{23} & \cdots & m_{2N} \\ 3 & t_{31} & t_{32} & t_{33} & \cdots & t_{3N} & m_{31} & m_{32} & m_{33} & \cdots & m_{3N} \\ \vdots & \vdots & \vdots & \vdots & \cdots & \vdots & \vdots & \vdots & \vdots & \cdots & \vdots \\ \vdots & \vdots & \vdots & \vdots & \cdots & \vdots & \vdots & \vdots & \vdots & \cdots & \vdots \\ P & t_{P1} & t_{P2} & t_{P3} & \cdots & t_{PN} & m_{P1} & m_{P2} & m_{P3} & \cdots & m_{PN} \end{bmatrix}$$

where *P* denotes the number of particles.

In a particular row, the first element denotes the particle number and the elements following it correspond to the position of that particle. Position includes the thicknesses of different layers followed by the respective material numbers. Similarly, the matrix (*VEL*) containing the initial velocities of all the parameters corresponding to each particle can be written as,

$$VEL = \begin{bmatrix} 1 & Vt_{11} & Vt_{12} & Vt_{13} & \cdots & Vt_{1N} & Vm_{11} & Vm_{12} & Vm_{13} & \cdots & Vm_{1N} \\ 2 & Vt_{21} & Vt_{22} & Vt_{23} & \cdots & Vt_{2N} & Vm_{21} & Vm_{22} & Vm_{23} & \cdots & Vm_{2N} \\ 3 & Vt_{31} & Vt_{32} & Vt_{33} & \cdots & Vt_{3N} & Vm_{31} & Vm_{32} & Vm_{33} & \cdots & Vm_{3N} \\ \vdots & \vdots & \vdots & \vdots & \cdots & \vdots & \vdots & \vdots & \vdots & \cdots & \vdots \\ \vdots & \vdots & \vdots & \vdots & \cdots & \vdots & \vdots & \vdots & \vdots & \cdots & \vdots \\ P & Vt_{P1} & Vt_{P2} & Vt_{P3} & \cdots & Vt_{PN} & Vm_{P1} & Vm_{P2} & Vm_{P3} & \cdots & Vm_{PN} \end{bmatrix}$$

Similar to *POP*, in a particular row, the first element denotes the particle number and the elements following it correspond to the velocities of all the parameters corresponding to that particle in the same order as defined in *POP*. The size of both *POP* and *VEL* matrices is $P \times (2N + 1)$.

Step#4: *Initialization of Pbest co-ordinates*

At the beginning of the algorithm, since initial positions are the only ones seen by all the particles, these positions have to be taken as each particle's respective *Pbest* co-ordinates. The *Pbest* co-ordinates can then be defined as,

$$Pbest\_coordinate = POP$$

Step#5: *Creation of a text file containing the material data corresponding to each particle*

For ease of data processing, a subroutine can be created to generate a text file containing the material data (frequency, real and imaginary parts of complex permittivity, real and imaginary parts of complex permeability, real and imaginary parts of phase constant) corresponding to the materials chosen for respective layers by each particle. In this book, the first column of the generated text file denotes the material numbers by which respective materials are referenced in the pre-defined material database.

Step#6: *Computation of fitness function corresponding to the initial position of each particle*

A subroutine implementing the expression for fitness function as presented in Equation (4.1) has to be called here to compute the value of the fitness function at the initial location of each particle. The fitness function values (FIT_VAL_i) associated with each particle are then stored in a $P \times 2$ matrix named *Particle_value*. At the beginning of the algorithm, since initial positions are the only ones seen by all the particles, the values at these positions are taken as each particle's respective *Pbest* values. Another $P \times 2$ matrix named *Pbest_value*, identical in structure to *Particle_value*, is used here to store the *Pbest* values corresponding to each particle.

$$Particle\_value = \begin{bmatrix} 1 & FIT\_VAL_1 \\ 2 & FIT\_VAL_2 \\ \vdots & \vdots \\ \vdots & \vdots \\ P & FIT\_VAL_P \end{bmatrix}$$

The first element in each row of *Particle_value* identifies the particle number (i) and the second element denotes the value of the fitness function corresponding to i^{th} particle (FIT_VAL_i). Further, at this step, $Pbest\_value = Particle\_value$.

Step#7: *Selection of Gbest*

The best value of fitness function amongst all the particles has to be selected from $Pbest\_value$ and stored in a variable named $Gbest\_value$. Since the summation of the overall recursive reflection and transmission coefficients over a range of frequencies and incident angles is taken as the fitness criteria, minimum value corresponds to the best value. The parameters of the particle with the minimum value of fitness function (chosen as $Gbest\_value$) are stored in a 1 × (2N + 1) matrix, named $Gbest\_coordinate$.

Step#8: *Evaluation of new positions and velocities*

The matrix VEL, containing the initial velocities of all the dimensions corresponding to each particle, has to be now updated using the following set of equations:

$$VEL(J,I) = W \times VEL(J,I) + PP + GG \tag{4.2}$$

where,

$$PP = c_1 \times rand(r_1) \times \big(pbest\_coordinate(J,I) - POP(J,I)\big)$$

$$GG = c_2 \times rand(r_2) \times \big(gbest\_coordinate(1,I) - POP(J,I)\big)$$

Here, a constant value of 1.49 has been assigned to both c_1 as well as c_2 and the inertial weight (W) is linearly decreased from 0.9 to 0.4 in each iteration. These values have been chosen in accordance with the criteria mentioned in (Robinson *et al.*, 2004). *rand*() generates random numbers less than 1 and greater than 0.

Once $VEL(J,I)$ is updated, the boundary conditions have to be checked and the final value can be written as,

$$VEL(J,I) = \min\big(VEL\_\max(J,I), \max\big(VEL\_\min(J,I), VEL(J,I)\big)\big) \tag{4.3}$$

The matrix POP, containing the initial positions of all particles, has to be now updated as,

$$POP(J,I) = POP(J,I) + VEL(J,I) \tag{4.4}$$

The updated positions have to be checked, and if found out-of-bound, the following boundary conditions, as discussed in Chapter 3, have to be applied:

If $POP(J,I) > DIM\_max(J,I)$ then,

$$POP(J,I) = DIM\_min(J,I) + rand(r_3) \times (DIM\_max(J,I) - DIM\_min(J,I))$$

If $POP(J,I) < DIM\_min(J,I)$ then,

$$POP(J,I) = DIM\_min(J,I) + rand(r_4) \times (DIM\_max(J,I) - DIM\_min(J,I))$$

Once the positions are updated, the new values for the fitness function have to be computed for each particle and stored in *Particle_value*. The updated fitness function values of all the particles have to be now compared with those stored in *Pbest_value* and appropriate replacements have to be made in accordance with the following condition:

If $Particle\_value(J,2) < Pbest\_vaue(J,2)$ then,

$$Pbest\_value(J,2) = Particle\_vaue(J,2)$$

$$Pbest\_coordinate(J,I) = POP(J,I)$$

Now, the best value of fitness function amongst all the particles has to be selected from the modified matrix *Pbest_value* and has to be stored in *Gbest_value*. The *Gbest_coordinate* has to be modified accordingly, too.

Step#9: Step#8 has to be repeated until the pre-defined number of iterations is completed. The final values stored in *Gbest_coordinate*, once all the iterations are completed, correspond to the required optimized solution.

An elaborate flowchart summarizing the entire algorithm is shown in Figure 4.1 .

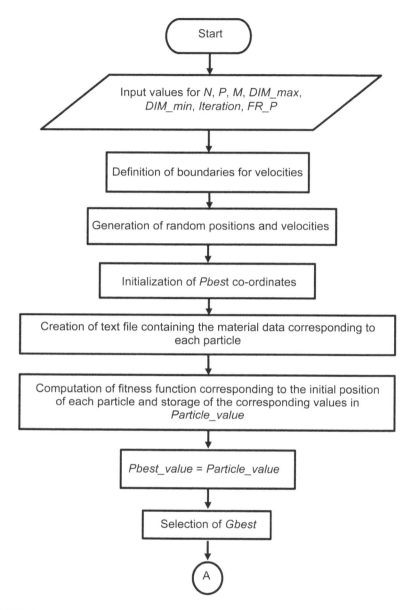

FIGURE 4.1
Flowchart for the optimal design of multilayered RAS using PSO.

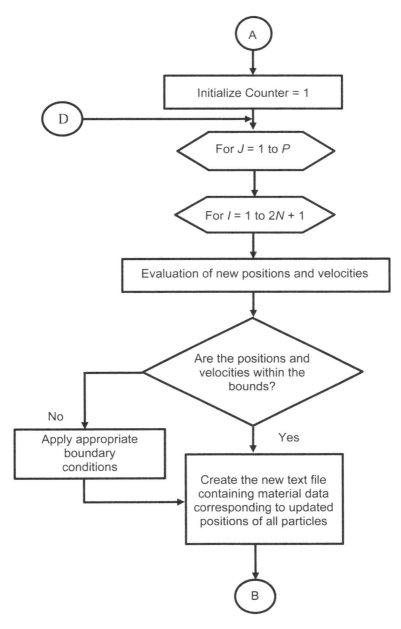

FIGURE 4.1
(Continued)

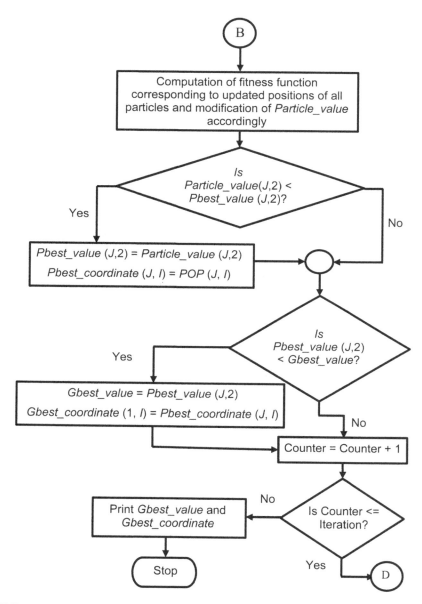

FIGURE 4.1
(Continued)

5

Validations and Performance Evaluation

The PSO-based algorithm for the optimization of multilayered RAS has been implemented using FORTRAN. The subroutine developed for the computation of reflection and transmission coefficients of multilayered RAS is validated in this chapter by comparing the results with those obtained from full wave simulation software (FEKO) as well as with those published in open domain. The superior performance of the indigenously developed code over FEKO is also established here by presenting the optimal design of a two-layered RAS using a pre-defined database of 31 materials. The constitutive parameters of all the materials considered in this chapter are summarized in the Appendix.

5.1 Validation of Developed Subroutine

A subroutine has been developed for the computation of overall recursive reflection and transmission coefficients of multilayered RAS, as per the formulation presented in Chapter 2. Let us consider the case of normal incidence first. The coefficients corresponding to normal incidence can be found by substituting $\theta_i = \theta_t = 0$ in the Equations (2.24) to (2.25) as well as (2.39) to (2.40), thereby simplifying all the cosine terms to 1. Now, consider a two-layered RAS (representative example) coated over 3.0 mm of CFRP with configuration as given below:

Top layer: Material#2 (Appendix); max thickness = 1.0 mm

Middle layer: Material#6 (Appendix); max thickness = 1.0 mm

Bottom layer: CFRP; thickness = 3.0 mm

The schematic diagram of the two-layered RAS is shown in Figure 5.1.

The subroutine developed for the computation of reflection and transmission coefficients has been used for the EM analysis of the two-layered RAS. The variation of return loss and insertion loss of the two-layered RAS *w.r.t.* frequency, for normal incidence case, evaluated using the developed subroutine, is shown in Figure 5.2 along with the results obtained from FEKO. The case of oblique incidence has also been studied for the considered RAS

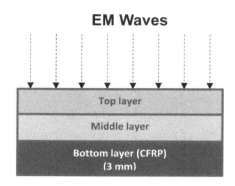

FIGURE 5.1
Schematic diagram of two-layered RAS.

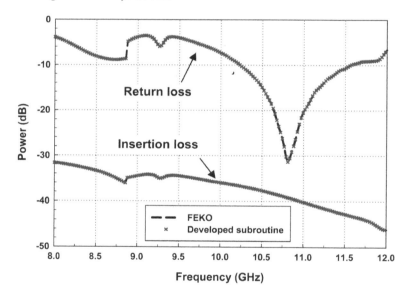

FIGURE 5.2
Variation of return loss and insertion loss of the two-layered RAS *w.r.t.* frequency (for normal incidence).

model. The variation of return loss and insertion loss *w.r.t.* frequency at different incident angles (θ) for TE polarized plane waves is shown in Figure 5.3. Similarly, the variation of return loss and insertion loss *w.r.t.* frequency at different incident angles for TM polarized plane waves is shown in Figure 5.4. The plots clearly indicate that the results obtained using developed subroutine are well matched with those obtained using full wave simulation software (FEKO), thereby establishing its authenticity.

The case of oblique incidence as presented in (Balanis, 2012) for a planar interface has also been considered here for validation of the subroutine for

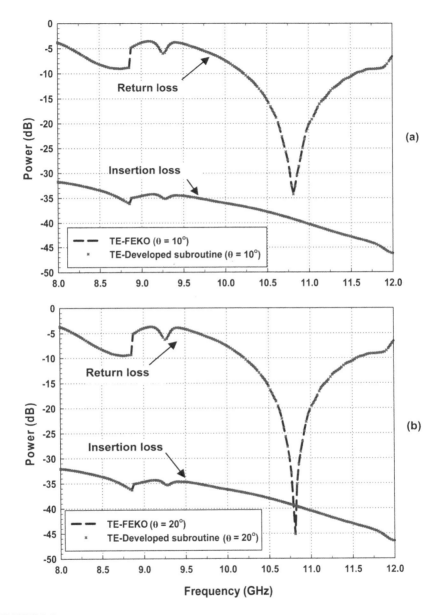

FIGURE 5.3
Comparison of power characteristics of two-layered RAS (for TE polarization) computed using developed subroutine and FEKO (a) θ = 10° (b) θ = 20° (c) θ = 30° (d) θ = 40° (e) θ = 50° (f) θ = 60° (g) θ = 70° (h) θ = 80° (i) consolidated.

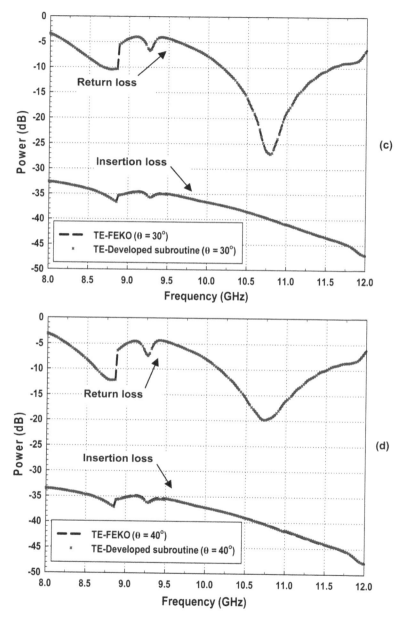

FIGURE 5.3
(Continued)

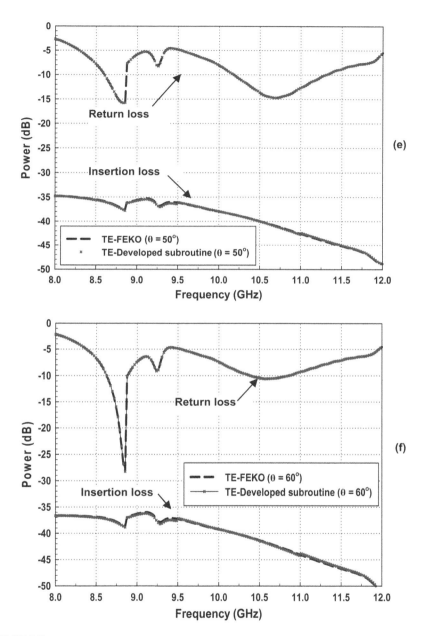

FIGURE 5.3
(Continued)

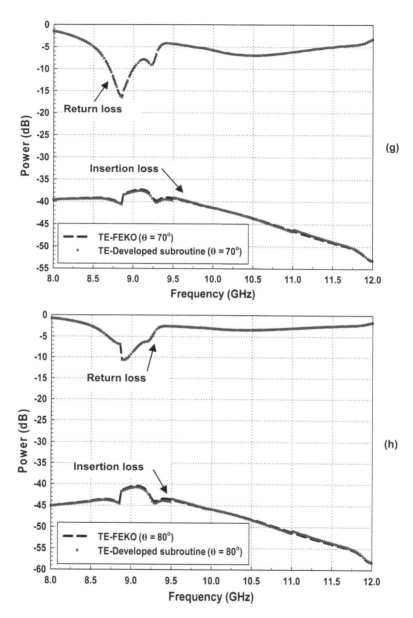

FIGURE 5.3
(Continued)

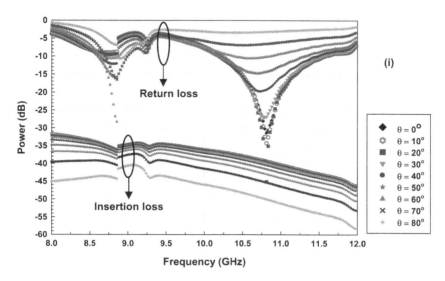

FIGURE 5.3
(Continued)

non-zero angle of incidence. The interface is shown in Figure 5.5. (ε_1, μ_1) and (ε_2, μ_2) denote the constitutive parameters corresponding to first medium and second medium, respectively. θ_i, θ_r and θ_t denote the angles made by incident ray, reflected ray and transmitted ray, respectively, with the normal to the interface.

The variation of reflection coefficient and transmission coefficient *w.r.t.* incident angle (for TE polarization) for different ratios of ε_1 and ε_2 are shown in Figure 5.6 and Figure 5.7, respectively. Similar plots for TM polarization are presented in Figure 5.8 and Figure 5.9, respectively. It is clear that the results obtained using the developed subroutine are perfectly matched *w.r.t.* those presented in (Balanis, 2012). The phenomenon of zero reflection for TM polarized plane waves at Brewster's angle can be clearly seen in Figure 5.8.

Once the subroutine has been validated, it has been incorporated into the PSO algorithm for the computation of overall recursive reflection and transmission coefficients as required in the evaluation of the following fitness function:

$$Fitness\ Function\ (F) = \sum_{i}^{n\_f} \sum_{j}^{n\_a} \left(\left| \tilde{R} \right|^2 + \left| \tilde{T} \right|^2 \right) \tag{5.1}$$

where n_f is the number of frequency points and n_a is the number of incident angles over which the RAS is to have optimum performance. The coefficients corresponding to TE polarization have to be considered for a TE polarized incident plane wave and vice versa.

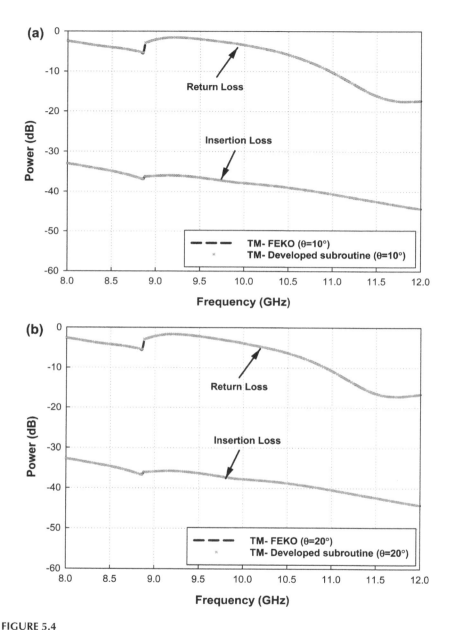

FIGURE 5.4
Comparison of power characteristics of two-layered RAS (for TM polarization) computed using developed subroutines and FEKO (a) $\theta = 10°$ (b) $\theta = 20°$ (c) $\theta = 30°$ (d) $\theta = 40°$ (e) $\theta = 50°$ (f) $\theta = 60°$ (g) $\theta = 70°$ (h) $\theta = 80°$ (i) consolidated.

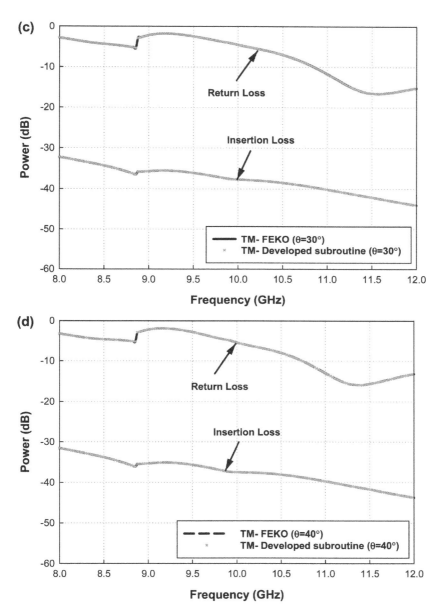

FIGURE 5.4
(Continued)

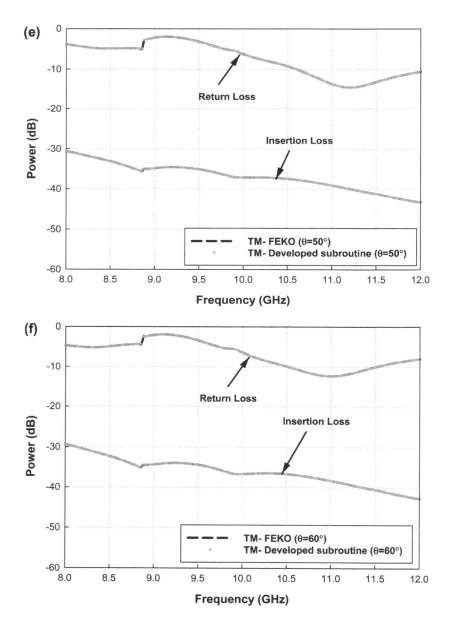

FIGURE 5.4
(Continued)

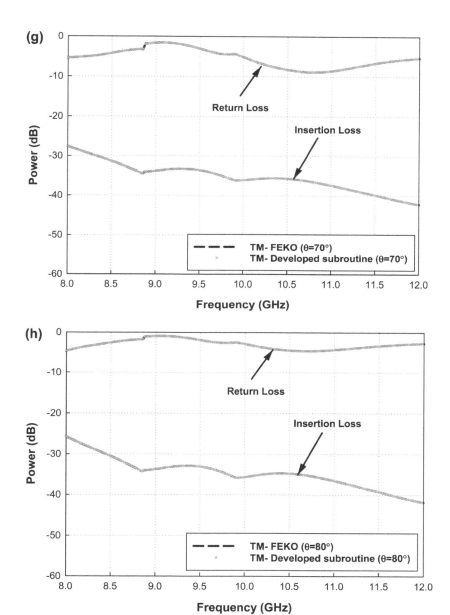

FIGURE 5.4
(Continued)

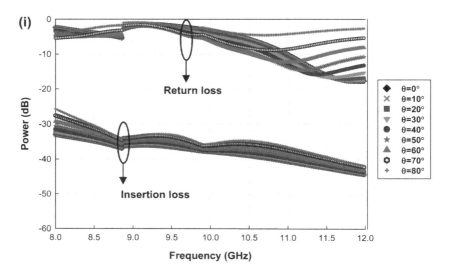

FIGURE 5.4
(Continued)

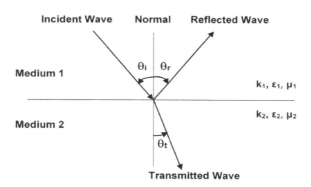

FIGURE 5.5
Oblique plane wave incidence on a planar interface.

5.2 Case Study for Performance Evaluation of Developed PSO-based Algorithm

Prior to the performance evaluation of developed PSO-based algorithm, a study has been carried out to analyze the variation in the value of fitness function *w.r.t.* the number of particles and number of iterations. This has been performed so that the parameters selected ensure the convergence of optimized solution. The two-layered RAS configuration presented in

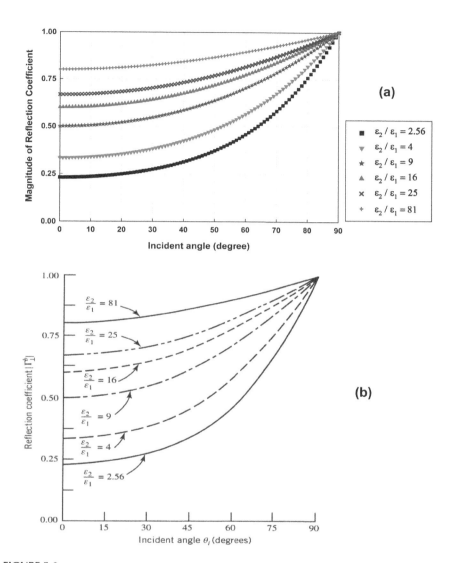

FIGURE 5.6
Variation of reflection coefficient *w.r.t.* incident angle for different ratios of ε_1 and ε_2 for TE polarization. (a) Computed using developed subroutine (b) Reference (Balanis, 2012).

Section 5.1 has been considered for this study with the maximum thickness for top and middle layers fixed at 2.0 mm. The variation in *Gbest_value* (the value of fitness function obtained using optimized material layer sequence and thickness profile) *w.r.t.* the number of iterations, fixing the number of particles as 30, is shown in Figure 5.10. In a similar manner, the variation in *Gbest_value w.r.t.* the number of particles, fixing the number of iterations as 30, is shown in Figure 5.11.

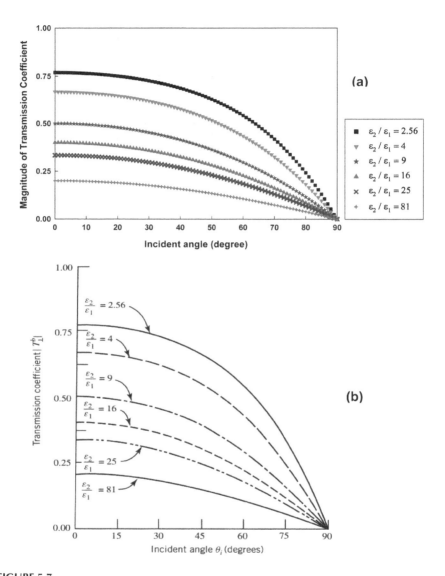

FIGURE 5.7
Variation of transmission coefficient *w.r.t.* incident angle for different ratios of ε_1 and ε_2 for TE polarization. (a) Computed using developed subroutine (b) Reference (Balanis, 2012).

Both the plots clearly suggest that 30 is a safe option for the number of iterations as well as the number of particles, as *Gbest_value* converges way before that. The same number has been recommended in (Robinson *et al.*, 2004) as their study suggested that most of the problems in electromagnetics converge well within 30 iterations and particles.

Once these parameters have been fixed, the indigenously developed code has been used for the optimal design of a two-layered RAS for superior

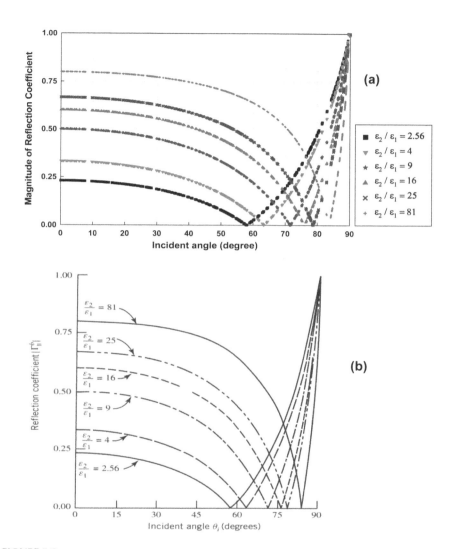

FIGURE 5.8
Variation of reflection coefficient *w.r.t.* incident angle for different ratios of ε_1 and ε_2 for TM polarization. (a) Computed using developed subroutine (b) Reference (Balanis, 2012).

absorption performance in X-band (8 GHz–12 GHz). A pre-defined database of 31 materials has been created and the constitutive parameters of all these materials at the center frequency (10 GHz) are summarized in the Appendix. The various parameters used while executing the code are summarized below:

Number of particles = 30
Number of iterations = 30

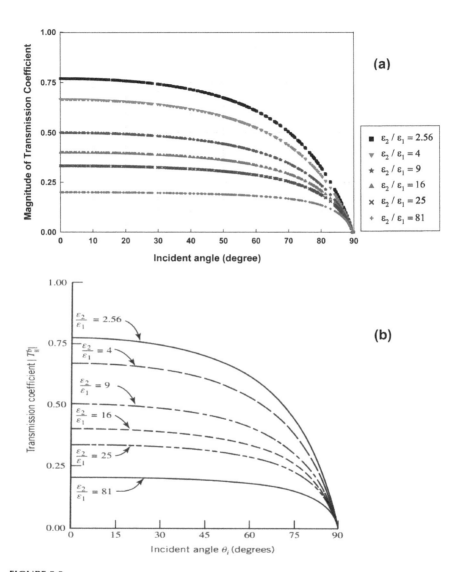

FIGURE 5.9
Variation of transmission coefficient *w.r.t.* incident angle for different ratios of ε_1 and ε_2 for TM polarization. (a) Computed using developed subroutine (b) Reference (Balanis, 2012).

Number of materials in database = 31
Maximum thickness of top layer = 1.5 mm
Maximum thickness of middle layer = 1.5 mm
Minimum thickness for all layers = 0.1 mm
Bottom layer: CFRP; thickness = 3.0 mm

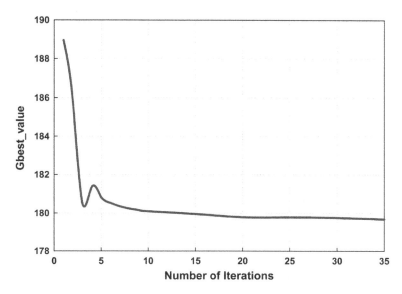

FIGURE 5.10
Variation in *Gbest_value w.r.t.* number of iterations fixing the number of particles as 30.

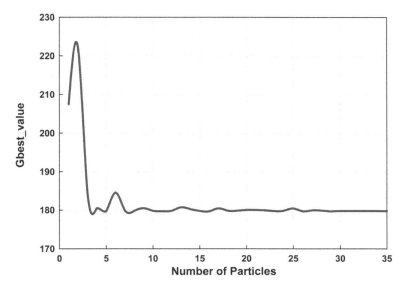

FIGURE 5.11
Variation in *Gbest_value w.r.t.* number of particles fixing the number of iterations as 30.

The final optimized material layer sequence as well as the thickness profile obtained on execution of the code is given below:

Top layer: Material#8 (Appendix); thickness = 1.36 mm

Middle layer: Material#22 (Appendix); thickness = 1.41 mm

Total thickness of RAS = 2.77 mm

The output from the code, including the variation of return loss, insertion loss and power absorption of the optimized RAS model *w.r.t.* frequency, is presented in Table 5.1. The graphical representation of the same is shown in Figure 5.12. It is clearly evident that the percentage of power absorbed is greater than 90% over the entire range of user-defined frequency band. The variation in the percentage of power absorbed *w.r.t.* frequency for different incident angles for both TE and TM polarizations is shown in Figure 5.13. The plots clearly indicate that the performance is well within the stipulated limits even at higher angles of incidence, thereby establishing the efficiency of the developed algorithm.

Now, let us analyze the difference in total simulation time between the indigenously developed PSO-based algorithm and the PSO-based optimization utility available in full wave simulation software (FEKO). Since FEKO doesn't provide the utility of automatic selection of appropriate materials for different layers, various material combinations have to be selected manually for the two-layered RAS model and their thicknesses have to be optimized separately in FEKO. With a database containing 31 materials, this implies that the number of combinations to be considered for a two-layered RAS model will be 31 × 31, resulting in 961 configurations. Therefore, the PSO-based optimization utility in FEKO has to be executed 961 times for all these possible material layer sequences in order to estimate the corresponding optimized thickness profiles. Once this is completed, the optimized results from these 961 evaluations have to be compiled and compared to arrive at the final optimized two-layered RAS design. The approximate simulation time in FEKO, for completing the PSO-based optimization of thickness profile of a two-layered RAS model over X-band (number of frequency points = 201; step size for thickness = 0.01 mm) is 2.5 hours. Dell precision tower 7810 workstation (Two Intel Xeon CPUs E5-2650 v4 @ 2.20 GHz; RAM: 256 GB) has been used for performing simulations. Therefore, the total time that FEKO would take for completing the optimization problem of a two-layered RAS model would be 961 × 2.5 ≈ 2,403 hours. i.e., approximately 100 days. On the other hand, the indigenously developed PSO-based code took less than a minute for the entire process, thereby clearly establishing its computational efficiency.

For illustration, the thickness profile of the optimized material layer sequence obtained from the developed code (Top layer: Material#8; Middle layer: Material#22), has been optimized in FEKO. The software took almost

TABLE 5.1

Power characteristics of optimized two-layered RAS model

Frequency (GHz)	Return loss (dB)	Return loss (%)	Insertion loss (dB)	Insertion loss (%)	Absorbed power (dB)	Absorbed power (%)
8.00	-13.94	4.04	-26.75	0.21	-0.19	95.75
8.02	-13.64	4.32	-26.73	0.21	-0.20	95.46
8.05	-13.48	4.48	-26.73	0.21	-0.21	95.30
8.07	-13.48	4.49	-26.77	0.21	-0.21	95.30
8.09	-13.60	4.37	-26.83	0.21	-0.20	95.43
8.11	-13.62	4.35	-26.86	0.21	-0.20	95.45
8.14	-13.96	4.02	-26.96	0.20	-0.19	95.78
8.16	-14.00	3.98	-27.01	0.20	-0.19	95.82
8.18	-14.03	3.96	-27.05	0.20	-0.18	95.85
8.20	-14.01	3.97	-27.08	0.20	-0.18	95.83
8.23	-13.69	4.27	-27.06	0.20	-0.20	95.53
8.25	-13.91	4.07	-27.14	0.19	-0.19	95.74
8.27	-13.70	4.27	-27.13	0.19	-0.20	95.54
8.29	-13.71	4.26	-27.17	0.19	-0.20	95.55
8.32	-13.47	4.50	-27.16	0.19	-0.21	95.30
8.34	-13.38	4.60	-27.17	0.19	-0.21	95.21
8.36	-13.55	4.42	-27.24	0.19	-0.20	95.39
8.38	-13.59	4.38	-27.29	0.19	-0.20	95.43
8.41	-13.61	4.35	-27.32	0.19	-0.20	95.46
8.43	-13.56	4.41	-27.34	0.18	-0.20	95.41
8.45	-13.69	4.27	-27.41	0.18	-0.20	95.54
8.47	-13.50	4.47	-27.42	0.18	-0.21	95.35
8.50	-13.61	4.35	-27.47	0.18	-0.20	95.47
8.52	-13.49	4.48	-27.49	0.18	-0.21	95.35
8.54	-13.58	4.38	-27.54	0.18	-0.20	95.44
8.56	-13.57	4.39	-27.57	0.17	-0.20	95.43
8.59	-13.71	4.26	-27.64	0.17	-0.20	95.57
8.61	-13.60	4.37	-27.66	0.17	-0.20	95.46
8.63	-13.28	4.69	-27.64	0.17	-0.22	95.13
8.65	-13.34	4.64	-27.68	0.17	-0.21	95.19
8.68	-13.28	4.70	-27.71	0.17	-0.22	95.13
8.70	-13.20	4.79	-27.74	0.17	-0.22	95.04
8.72	-13.14	4.85	-27.76	0.17	-0.22	94.98
8.74	-13.14	4.85	-27.80	0.17	-0.22	94.98
8.77	-13.24	4.74	-27.86	0.16	-0.22	95.09
8.79	-13.40	4.57	-27.93	0.16	-0.21	95.27
8.81	-13.39	4.58	-27.97	0.16	-0.21	95.26
8.83	-13.56	4.41	-28.05	0.16	-0.20	95.44
8.86	-13.71	4.26	-28.10	0.15	-0.20	95.59

(Continued)

TABLE 5.1 (CONTINUED)

Power characteristics of optimized two-layered RAS model

Frequency (GHz)	Return loss (dB)	Return loss (%)	Insertion loss (dB)	Insertion loss (%)	Absorbed power (dB)	Absorbed power (%)
8.88	-13.81	4.16	-28.15	0.15	-0.19	95.68
8.90	-13.93	4.05	-28.22	0.15	-0.19	95.80
8.92	-13.93	4.05	-28.26	0.15	-0.19	95.80
8.95	-13.97	4.01	-28.30	0.15	-0.18	95.84
8.97	-14.04	3.95	-28.36	0.15	-0.18	95.91
8.99	-14.14	3.86	-28.42	0.14	-0.18	96.00
9.01	-14.29	3.73	-28.48	0.14	-0.17	96.13
9.04	-14.45	3.59	-28.56	0.14	-0.16	96.27
9.06	-14.53	3.52	-28.61	0.14	-0.16	96.34
9.08	-14.56	3.50	-28.65	0.14	-0.16	96.37
9.10	-14.65	3.43	-28.70	0.13	-0.16	96.44
9.13	-14.77	3.33	-28.78	0.13	-0.15	96.54
9.15	-14.89	3.24	-28.84	0.13	-0.15	96.63
9.17	-14.97	3.19	-28.89	0.13	-0.15	96.68
9.19	-15.01	3.16	-28.94	0.13	-0.15	96.71
9.22	-15.07	3.11	-29.00	0.13	-0.14	96.77
9.24	-15.29	2.96	-29.07	0.12	-0.14	96.92
9.26	-15.38	2.90	-29.12	0.12	-0.13	96.98
9.28	-15.44	2.86	-29.17	0.12	-0.13	97.02
9.31	-15.53	2.80	-29.22	0.12	-0.13	97.08
9.33	-15.51	2.81	-29.25	0.12	-0.13	97.07
9.35	-15.66	2.72	-29.31	0.12	-0.12	97.17
9.37	-15.61	2.75	-29.34	0.12	-0.13	97.13
9.40	-15.62	2.74	-29.38	0.12	-0.13	97.14
9.42	-15.81	2.63	-29.46	0.11	-0.12	97.26
9.44	-15.94	2.55	-29.52	0.11	-0.12	97.34
9.46	-16.01	2.51	-29.57	0.11	-0.12	97.38
9.49	-15.99	2.52	-29.61	0.11	-0.12	97.38
9.51	-16.00	2.51	-29.66	0.11	-0.12	97.38
9.53	-16.03	2.49	-29.70	0.11	-0.11	97.40
9.55	-16.09	2.46	-29.75	0.11	-0.11	97.43
9.58	-16.20	2.40	-29.81	0.10	-0.11	97.50
9.60	-16.20	2.40	-29.85	0.10	-0.11	97.50
9.62	-16.27	2.36	-29.89	0.10	-0.11	97.54
9.64	-16.31	2.34	-29.95	0.10	-0.11	97.56
9.67	-16.39	2.30	-29.99	0.10	-0.11	97.60
9.69	-16.33	2.33	-30.02	0.10	-0.11	97.57
9.71	-16.52	2.23	-30.09	0.10	-0.10	97.68
9.73	-16.41	2.29	-30.11	0.10	-0.10	97.62
9.76	-16.65	2.16	-30.18	0.10	-0.10	97.74

(Continued)

TABLE 5.1 (CONTINUED)

Power characteristics of optimized two-layered RAS model

Frequency (GHz)	Return loss (dB)	Return loss (%)	Insertion loss (dB)	Insertion loss (%)	Absorbed power (dB)	Absorbed power (%)
9.78	-16.64	2.17	-30.22	0.10	-0.10	97.74
9.80	-16.72	2.13	-30.27	0.09	-0.10	97.78
9.82	-16.81	2.08	-30.32	0.09	-0.10	97.82
9.89	-17.09	1.95	-30.48	0.09	-0.09	97.96
9.91	-17.18	1.91	-30.53	0.09	-0.09	98.00
9.94	-17.25	1.88	-30.58	0.09	-0.09	98.03
9.96	-17.38	1.83	-30.64	0.09	-0.08	98.08
9.98	-17.49	1.78	-30.69	0.09	-0.08	98.13
10.00	-17.48	1.79	-30.73	0.08	-0.08	98.13
10.03	-17.59	1.74	-30.78	0.08	-0.08	98.18
10.05	-17.65	1.72	-30.84	0.08	-0.08	98.20
10.07	-17.75	1.68	-30.88	0.08	-0.08	98.24
10.09	-17.99	1.59	-30.96	0.08	-0.07	98.33
10.12	-17.83	1.65	-30.99	0.08	-0.08	98.27
10.14	-17.91	1.62	-31.04	0.08	-0.07	98.30
10.16	-18.04	1.57	-31.09	0.08	-0.07	98.35
10.18	-18.12	1.54	-31.15	0.08	-0.07	98.38
10.21	-18.13	1.54	-31.18	0.08	-0.07	98.39
10.23	-18.26	1.49	-31.24	0.08	-0.07	98.43
10.25	-18.40	1.45	-31.30	0.07	-0.07	98.48
10.27	-18.48	1.42	-31.36	0.07	-0.07	98.51
10.30	-18.57	1.39	-31.40	0.07	-0.06	98.54
10.32	-18.80	1.32	-31.47	0.07	-0.06	98.61
10.34	-19.06	1.24	-31.55	0.07	-0.06	98.69
10.36	-19.27	1.18	-31.62	0.07	-0.05	98.75
10.39	-19.47	1.13	-31.68	0.07	-0.05	98.80
10.50	-20.14	0.97	-31.98	0.06	-0.05	98.97
10.52	-20.20	0.96	-32.03	0.06	-0.04	98.98
10.59	-20.59	0.87	-32.20	0.06	-0.04	99.07
10.61	-20.60	0.87	-32.25	0.06	-0.04	99.07
10.63	-20.86	0.82	-32.33	0.06	-0.04	99.12
10.66	-20.89	0.81	-32.38	0.06	-0.04	99.13
10.68	-21.04	0.79	-32.44	0.06	-0.04	99.16
10.70	-21.10	0.78	-32.49	0.06	-0.04	99.17
10.72	-21.20	0.76	-32.55	0.06	-0.04	99.19
10.75	-21.27	0.75	-32.61	0.05	-0.03	99.20
10.77	-21.55	0.70	-32.68	0.05	-0.03	99.25
10.79	-21.73	0.67	-32.75	0.05	-0.03	99.28
10.81	-21.82	0.66	-32.80	0.05	-0.03	99.29
10.84	-21.94	0.64	-32.86	0.05	-0.03	99.31

(*Continued*)

TABLE 5.1 (CONTINUED)

Power characteristics of optimized two-layered RAS model

Frequency (GHz)	Return loss (dB)	Return loss (%)	Insertion loss (dB)	Insertion loss (%)	Absorbed power (dB)	Absorbed power (%)
10.86	-22.21	0.60	-32.93	0.05	-0.03	99.35
10.88	-22.30	0.59	-32.99	0.05	-0.03	99.36
10.90	-22.57	0.55	-33.07	0.05	-0.03	99.40
10.93	-22.86	0.52	-33.15	0.05	-0.02	99.43
10.95	-23.21	0.48	-33.24	0.05	-0.02	99.47
10.97	-23.24	0.47	-33.30	0.05	-0.02	99.48
10.99	-23.30	0.47	-33.36	0.05	-0.02	99.49
11.02	-23.38	0.46	-33.42	0.05	-0.02	99.50
11.04	-23.36	0.46	-33.48	0.04	-0.02	99.49
11.06	-23.37	0.46	-33.54	0.04	-0.02	99.50
11.08	-23.47	0.45	-33.61	0.04	-0.02	99.51
11.11	-23.54	0.44	-33.67	0.04	-0.02	99.51
11.13	-23.64	0.43	-33.73	0.04	-0.02	99.53
11.15	-23.74	0.42	-33.79	0.04	-0.02	99.54
11.17	-23.86	0.41	-33.86	0.04	-0.02	99.55
11.20	-23.89	0.41	-33.93	0.04	-0.02	99.55
11.22	-23.90	0.41	-33.99	0.04	-0.02	99.55
11.24	-23.89	0.41	-34.05	0.04	-0.02	99.55
11.26	-24.02	0.40	-34.11	0.04	-0.02	99.56
11.29	-23.95	0.40	-34.16	0.04	-0.02	99.56
11.31	-24.12	0.39	-34.25	0.04	-0.02	99.57
11.38	-23.89	0.41	-34.45	0.04	-0.02	99.56
11.40	-23.80	0.42	-34.52	0.04	-0.02	99.55
11.42	-23.65	0.43	-34.58	0.03	-0.02	99.53
11.44	-23.60	0.44	-34.65	0.03	-0.02	99.53
11.47	-23.64	0.43	-34.71	0.03	-0.02	99.53
11.49	-23.58	0.44	-34.79	0.03	-0.02	99.53
11.51	-23.46	0.45	-34.85	0.03	-0.02	99.52
11.53	-23.42	0.46	-34.92	0.03	-0.02	99.51
11.56	-23.33	0.46	-34.99	0.03	-0.02	99.50
11.58	-23.29	0.47	-35.06	0.03	-0.02	99.50
11.60	-23.14	0.49	-35.13	0.03	-0.02	99.48
11.62	-23.09	0.49	-35.20	0.03	-0.02	99.48
11.65	-23.01	0.50	-35.27	0.03	-0.02	99.47
11.67	-22.84	0.52	-35.34	0.03	-0.02	99.45
11.69	-22.69	0.54	-35.40	0.03	-0.02	99.43
11.71	-22.60	0.55	-35.47	0.03	-0.03	99.42
11.74	-22.43	0.57	-35.54	0.03	-0.03	99.40
11.76	-22.26	0.59	-35.59	0.03	-0.03	99.38
11.78	-22.10	0.62	-35.65	0.03	-0.03	99.36

(Continued)

TABLE 5.1 (CONTINUED)

Power characteristics of optimized two-layered RAS model

Frequency (GHz)	Return loss (dB)	Return loss (%)	Insertion loss (dB)	Insertion loss (%)	Absorbed power (dB)	Absorbed power (%)
11.80	-21.96	0.64	-35.71	0.03	-0.03	99.34
11.83	-21.81	0.66	-35.76	0.03	-0.03	99.31
11.85	-21.69	0.68	-35.82	0.03	-0.03	99.30
11.87	-21.59	0.69	-35.88	0.03	-0.03	99.28
11.94	-21.20	0.76	-36.06	0.02	-0.03	99.22
11.96	-21.13	0.77	-36.13	0.02	-0.03	99.21
11.98	-20.97	0.80	-36.18	0.02	-0.04	99.18
12.01	-20.86	0.82	-36.27	0.02	-0.04	99.16
12.03	-20.78	0.84	-36.34	0.02	-0.04	99.14
12.05	-20.61	0.87	-36.40	0.02	-0.04	99.11
12.07	-20.41	0.91	-36.47	0.02	-0.04	99.07
12.10	-20.27	0.94	-36.54	0.02	-0.04	99.04
12.12	-20.14	0.97	-36.61	0.02	-0.04	99.01
12.14	-20.00	1.00	-36.68	0.02	-0.04	98.98
12.16	-19.85	1.03	-36.77	0.02	-0.05	98.94
12.19	-19.76	1.06	-36.83	0.02	-0.05	98.92
12.21	-19.61	1.09	-36.89	0.02	-0.05	98.89
12.23	-19.42	1.14	-36.94	0.02	-0.05	98.84
12.25	-19.33	1.17	-37.02	0.02	-0.05	98.81
12.28	-19.19	1.21	-37.09	0.02	-0.05	98.77
12.30	-19.10	1.23	-37.16	0.02	-0.05	98.75
12.32	-19.08	1.24	-37.23	0.02	-0.05	98.74
12.34	-18.98	1.26	-37.27	0.02	-0.06	98.72
12.37	-18.92	1.28	-37.32	0.02	-0.06	98.70
12.39	-18.85	1.30	-37.37	0.02	-0.06	98.68
12.41	-18.78	1.32	-37.43	0.02	-0.06	98.66
12.43	-18.66	1.36	-37.49	0.02	-0.06	98.62
12.46	-18.54	1.40	-37.55	0.02	-0.06	98.58
12.48	-18.45	1.43	-37.61	0.02	-0.06	98.55
12.50	-18.37	1.46	-37.64	0.02	-0.06	98.53

2.5 hours for the optimization process as the models corresponding to each parametric change were solved separately by full wave analysis. The optimum total thickness of RAS as recommended by FEKO has been found to be 2.72 mm. The variation in the percentage of power absorbed *w.r.t.* frequency for both the models is shown in Figure 5.14. It clearly shows that the performance is almost the same for both the cases, but the code achieved it within a much shorter span of time. The performance of both the platforms *w.r.t.* total thickness of RAS and simulation time is summarized in Table 5.2.

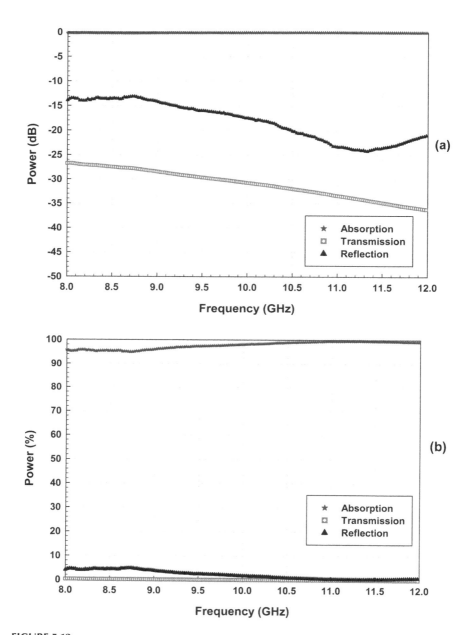

FIGURE 5.12
Variation in power characteristics of optimized two-layered RAS model *w.r.t.* frequency (a) in
dB (b) in %.

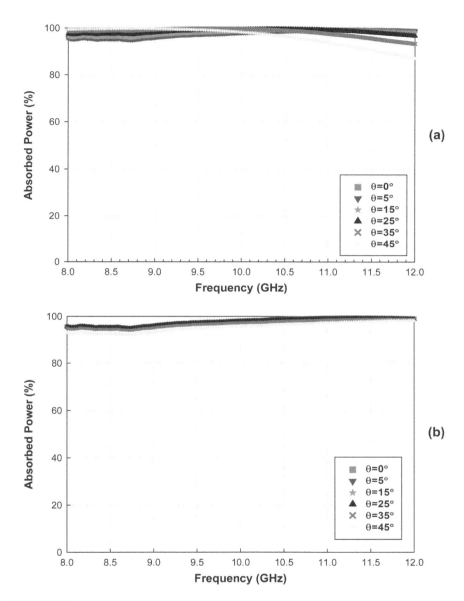

FIGURE 5.13
Variation in the power absorption characteristics of the optimized two-layered RAS model *w.r.t.* frequency for different incident angles (a) TE polarization (b) TM polarization.

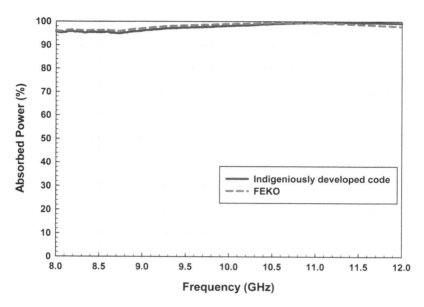

FIGURE 5.14
Comparison between the performance of PSO-based optimization utility available in FEKO and the indigenously developed PSO-based algorithm for the optimal design of RAS.

TABLE 5.2

Comparison of performance between developed code and full wave simulation software (FEKO)

Performance parameter	Indigenously developed code	FEKO
Total thickness of RAS	2.77 mm (Simulation performed using the material database containing 31 materials)	2.72 mm (Simulation performed only for the optimized material layer sequence obtained from the developed code)
Total simulation time	< 1 minute	2,403 hours (Approximately calculated as $31 \times 31 \times 2.5$)

5.3 Conclusion

In this exceedingly technology-driven industrial scenario, microwave absorbers are being used extensively for various applications in both military and civilian domains. In the military arena, they are used in stealth vehicles to reduce the visibility from rival radar systems. On the other hand, in civilian sector, they are mainly used to reduce the electromagnetic interference (EMI) between various microwave/electronic components. The primary aim of a microwave absorber is to achieve maximum power absorption in a broad range of frequencies for any angle of incidence and arbitrary polarization. Amongst the different kinds of microwave absorbers, including FSS-based absorbers, multilayered radar absorbing structures have gained exceptional research interest recently due to the rapid advances in the field of material science and technology. Furthermore, they are widely used in stealth applications due to their broad bandwidth and reduced weight penalty. The selection of suitable material for each layer and the optimization of the corresponding thickness profile are the two important factors which determine the performance of multilayered RAS. Although commercially available full wave simulation software provide algorithms (like particle swarm optimization, genetic algorithm, etc.), for optimization of thickness, they do not have options for optimizing the position of a particular material inside the multilayered RAS configuration. In this regard, the present book has described the development of an efficient algorithm, based on PSO, for the material selection as well as optimization of thickness profile of multilayered RAS models considering both normal as well as oblique incidence cases. The theoretical formulation for the computation of overall recursive reflection and transmission coefficients of multilayered RAS configurations has been elaborately explained with sufficient validations. The underlying physical interpretations, key parameters involved and the step-by-step procedure for implementing PSO in the context of design and optimization of multilayered RAS have also been discussed in detail. Further, the superior performance of the indigenously developed FORTRAN-based code in comparison with full wave simulation software (FEKO), *w.r.t.* simulation time, has been established by presenting the optimal design of a two-layered RAS using a pre-defined database of 31 materials.

Appendix: Database of Materials Used for the Optimal Design of Multilayered RAS

The constitutive parameters of all the materials considered in this book at the center frequency (10 GHz) are summarized here. The EM parameters include the real and imaginary parts of relative permittivity as well as relative permeability. Further, Material#32 corresponds to CFRP, which is automatically taken as the base material for the multilayered RAS model.

TABLE A.1

Constitutive parameters of materials at 10 GHz

Material no.	ε_r'	ε_r''	μ_r'	μ_r''
1	6.50	0.21	1.00	0.15
2	5.00	0.10	1.00	0.01
3	8.0	1.50	0.90	0.03
4	10.00	1.41	0.85	0.14
5	6.90	0.70	0.80	0.01
6	0.16	0.60	2.70	1.0
7	1.80	6.5	0.01	0.30
8	9.00	0.20	0.50	0.30
9	8.00	0.16	0.50	0.08
10	6.50	1.20	0.27	0.31
11	9.50	0.75	0.65	0.20
12	1.0	0.40	0.60	0.22
13	13.00	10.00	0.55	0.20
14	6.50	0.19	1.20	0.15
15	10.00	2.40	0.92	0.16
16	9.58	0.17	0.351	0.195
17	6.81	2.00	1.0	0.03
18	5.80	0.10	0.50	0.30
19	7.49	0.250	0.361	0.31
20	7.20	0.261	0.361	0.349
21	6.810	0.210	0.400	0.319
22	5.410	3.169	0.981	0.050
23	6.29	0.30	0.60	0.10
24	1.20	1.1801	0.850	0.09
25	3.80	1.39	1.03	0.50

(*Continued*)

TABLE A.1 (CONTINUED)

Constitutive parameters of materials at 10 GHz

Material no.	ε_r'	ε_r''	μ_r'	μ_r''
26	4.131	0.121	0.810	0.180
27	0.4	0.100	1.651	0.191
28	3.699	0.3214	0.969	0.0020
29	3.71	0.129	1.001	0.0041
30	4.301	0.459	0.9985	0.0080
31	5.599	1.2499	1.00	0.0401
32	115.00	75.500	0.0698	0.100

References

Asi, M., and N. I. Dib, "Design of multilayer microwave broadband absorbers using central force optimization," *Progress in Electromagnetics Research*, vol. 26, pp. 101–113, September 2010.

Balanis, C. A., *Advanced Engineering Electromagnetics*, John Wiley & Sons Inc., ISBN: 978-0-470-58948-9, 1018 p., 2012.

Chakravarty, S., R. Mittra, and N. R. Williams, "Application of a microgenetic algorithm (MGA) to the design of broadband microwave absorbers using multiple frequency selective surface screens buried in dielectrics," *IEEE Transactions on Antennas and Propagation*, vol. 50, no. 3, pp. 284–296, March 2002.

Chamaani, S., S. A. Mirtaheri, M. Teshnehlab, M. A. Shoorehdeli, and V. Seydi, "Modified multi-objective particle swarm optimization for electromagnetic absorber design," *Progress in Electromagnetics Research*, vol. 79, pp. 353–366, 2008.

Chew, W. C., *Waves and Fields in Inhomogeneous Media*, IEEE Publications, ISBN: 0-7803-4749-8, pp. 45–53, 1995.

Cui, S., D. S. Weile, and J. L. Volakis, "Novel planar electromagnetic absorber designs using genetic algorithms," *IEEE Transactions on Antennas and Propagation*, vol. 54, no. 6, pp. 1811–1817, June 2006.

Goudos, S. K., "Design of microwave broadband absorbers using a self-adaptive differential evolution algorithm," *International Journal of RF and Microwave Computer-Aided Engineering*, vol. 19, no. 3, pp. 364–372, December 2008.

Goudos, S. K., and J. N. Sahalos, "Design of broadband radar absorbing materials using particle swarm optimization," *Proceedings of EMC Europe 2006 International Symposium on Electromagnetic Compatibility*, Barcelona, Spain, September 4–8, 2006, pp. 1111–1116.

Hassan, R., B. Cohanim, O. D. Weck, and G. Venter, "A comparison of particle swarm optimization and genetic algorithm," *Proceedings of the 1st AIAA Multidisciplinary Design Optimization Specialist Conference*, Texas, United States, April 18–21, 2005.

Jiang, L., J. Cui, L. Shi, and X. Li, "Pareto optimal design of multi-layer microwave absorbers for wide-angle incidence using genetic algorithms," *IET Microwaves, Antennas and Propagation*, vol. 3, no. 4, pp. 572–579, June 2009.

Kern, D. J., and D. H. Werner, "A genetic algorithm approach to the design of ultra-thin electromagnetic bandgap absorbers," *Microwave and Optical Technology Letters*, vol. 38, no. 1, pp. 61–64, July 2003.

Knott, E. F., J. F. Shaeffer, and M. T. Tuley, *Radar Cross Section*, Artech House, ISBN: 978–1891121258, 477 p., 1985.

Michielssen, E., J. M. Sajer, S. Ranjithan, and R. Mittra, "Design of lightweight, broadband microwave absorbers using genetic algorithms," *IEEE Transactions on Microwave Theory and Techniques*, vol. 41, no. 6, pp. 1024–1031, June 1993.

Perini, J., and L. S. Cohen, "Design of broad-band radar-absorbing materials for large angles of incidence," *IEEE Transactions on Electromagnetic Compatibility*, vol. 35, no. 2, pp. 223–230, May 1993.

Pozar, D. M., *Microwave Engineering*, John Wiley & Sons Inc., ISBN: 9971-51-263-7, pp. 39–43, 1998.

Ranjan, P., "Wide-angle polarization independent multilayer microwave absorber using wind driven optimization technique," *International Journal of Applied Engineering Research*, vol. 12, no. 19, pp. 8016–8025, 2017.

Robinson, J., and Y. R. Samii, "Particle swarm optimization in electromagnetics," *IEEE Transactions on Antennas and Propagation*, vol. 52, no. 2, pp. 397–407, February 2004.

Roy, S., S. D. Roy, J. Tewary, A. Mahanti, and G. K. Mahanti, "Particle swarm optimization for optimal design of broadband multilayer microwave absorber for wide angle of incidence," *Progress in Electromagnetics Research B*, vol. 62, pp. 121–135, 2015.

Singh, H., D. D. J. Ebison, H. S. Rawat, and R. George, *Fundamentals of EM Design of Radar Absorbing Structures (RAS)*, Springer Briefs in Applied Sciences and Technology, ISBN 978-981-10-5079-4, 55 p., 2017.

Toktas, A., D. Ustun, E. Yigit, K. Sabanci, and M. Tekbas, "Optimally synthesizing multilayer radar absorbing material (RAM) using artificial bee colony algorithm," *Proceedings of 2018 XXIIIrd International Seminar/Workshop on Direct and Inverse Problems of Electromagnetic and Acoustic Wave Theory (DIPED)*, Tbilisi, September 24–27, 2018, pp. 237–241.

Toktas, A., D. Ustun, and M. Tekbas, "Multi-objective design of multi-layer radar absorber using surrogate-based optimization," *IEEE Transactions on Microwave Theory and Techniques*, vol. 67, no. 8, pp. 3318–3329, August 2019.

Vinoy, K. J., and R. M. Jha, *Radar Absorbing Materials: From Theory to Design and Characterization*, Kluwer Academic Publishers, ISBN 978-1-4613-8065-8, 190 p., 1996.

Yigit, E., and H. Duysak, "Determination of optimal layer sequence and thickness for broadband multilayer absorber design using double-stage artificial bee colony algorithm," *IEEE Transactions on Microwave Theory and Techniques*, vol. 67, no. 8, pp. 3306–3317, August 2019.

Author Index

Subject Index

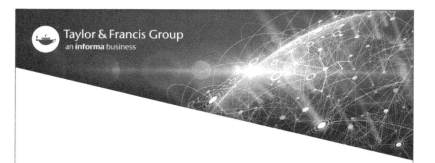